Digital Holography and Wavefront Sensing

Ulf Schnars · Claas Falldorf
John Watson · Werner Jüptner

Digital Holography and Wavefront Sensing

Principles, Techniques and Applications

Second Edition

 Springer

Ulf Schnars
Hagen
Germany

Claas Falldorf
Bremer Institut für angewandte
Strahltechnik (BIAS)
Bremen
Germany

John Watson
School of Engineering
University of Aberdeen
Aberdeen
Scotland, UK

Werner Jüptner
Ritterhude
Germany

ISBN 978-3-662-44692-8   ISBN 978-3-662-44693-5   (eBook)
DOI 10.1007/978-3-662-44693-5

Library of Congress Control Number: 2014949296

Springer Heidelberg New York Dordrecht London

Preface to the Second Edition

Life always bursts the boundaries of formulas

Antoine de Saint Exupéry

As we sat down to consider writing a new edition of *Digital Holography*, we, the original authors (U. Schnars and W. Jüptner), asked ourselves if the field had advanced sufficiently with enough new and novel developments to merit a second edition. The answer was an overwhelming YES and came from seeing the profound developments and scale of applications to which digital holography, and in the wider context 3D imaging technologies in general, are now being routinely applied. In the intervening years, the evolution of digital holography has been, both, extensive and dramatic.

Some of the areas in which we have seen considerable advances and application include computational wave field sensing and digital holographic microscopy, with a huge number of papers being published in these and related fields. To reflect these advances adequately in our book and to broaden its scope, we invited Claas Falldorf (BIAS) and John Watson (University of Aberdeen) to join us as co-authors. Claas works actively in wave field sensing using computational methods such as phase retrieval or computational shear interferometry. John is an international expert in digital holographic microscopy and, particularly, to underwater holography of aquatic organisms and particles; and also 3DTV and related fields. Both are ideal partners to support the approach and philosophy of the new edition.

Accordingly, this second edition has been significantly revised and enlarged. We have extended the chapter on Digital Holographic Microscopy to incorporate new sections on particle sizing, particle image velocimetry and underwater holography. A new chapter now deals comprehensively and extensively with computational wave field sensing. These techniques represent a fascinating alternative to standard interferometry and Digital Holography. They enable wave field sensing without the requirement of a particular reference wave, thus allowing the use of low brilliance light sources and even liquid-crystal displays (LCD) for interferometric applications. We believe that, in the coming years, computational wave field sensing will prove to be an excellent complement to Digital Holography to determine the full complex amplitude of wave fields.

All the authors wish to thank colleagues past and present (too numerous to mention) with whom they have worked over the years. As with the first edition, several pictures and figures in this book originate from common publications with other colleagues and we thank them for permission to describe their work and to use their pictures. All of our co-workers are gratefully acknowledged.

Bremen, May 2014   Ulf Schnars
Aberdeen   Claas Falldorf
John Watson
Werner Jüptner

Preface to the First Edition

*Sag' ich zum Augenblicke verweile doch, Du bist
so schön*

J.W.v. Goethe, "Faust"

An old dream of mankind and a sign of culture is the conservation of moments by taking an image of the world around. Pictures accompany the development of mankind. However, a picture is the two-dimensional projection of the three-dimensional world. The perspective—recognized in Europe in the Middle Ages—was a first approach to overcome the difficulties of imaging close to reality. It took up to the twentieth century to develop a real three-dimensional imaging: Gabor invented holography in 1948. Yet still one thing was missing: the phase of the object wave could be reconstructed optically but not be measured directly. The last huge step to the complete access of the object wave was Digital Holography. By Digital Holography the intensity and the phase of electromagnetic wave fields can be measured, stored, transmitted, applied to simulations and manipulated in the computer: An exciting new tool for the handling of light.

We started our work in the field of Digital Holography in 1990. Our motivation mainly came from Holographic Interferometry, a method used with success for precise measurement of deformation and shape of opaque bodies or refractive index variations within transparent media. A major drawback of classical HI using photographic plates was the costly process of film development. Even thermoplastic films used as recording medium did not solve the hologram development problem successfully. On the other hand the Electronic Speckle Pattern Interferometry (ESPI) and its derivate digital shearography reached a degree mature for applications in industry. Yet, with these speckle techniques the recorded images are only correlated and not reconstructed as for HI. Characteristic features of holography like the possibility to refocus on other object planes in the reconstruction process are not possible with speckle metrology.

Our idea was to transfer all methods of classical HI using photographic plates to Digital Holography. Surprisingly, we discovered that Digital Holography offers more possibilities than classical HI: The wavefronts can be manipulated in the numerical reconstruction process, enabling operations not possible in optical holography. Especially the interference phase can be calculated directly from the holograms, without evaluation of an interference pattern.

The efficiency of Digital Holography depends strongly on the resolution of the electronic target used to record the holograms. When we made our first experiments in the 1990s of the last century, Charged Coupled Devices began to replace analogue sensors in cameras. The resolution of commercially available cameras was quite low, about some hundred pixels per line, and the output signal of cameras already equipped with CCDs was still analogue. In those days, digital sampling of camera images and running of routines for numerical hologram reconstruction was only possible on special digital image processing hardware and not, as today, on ordinary PCs. The reconstruction of a hologram digitized with 512×512 pixels took about half an hour in 1991 on a Digital Image Processing unit developed at BIAS especially for optical metrology purposes. Nevertheless we made our first experiments with these types of cameras. Today, numerical reconstruction of holograms with 1 million pixel is possible nearly in real time on state-of-the-art PCs.

Then, fully digital CCD cameras with 1 million pixels and smaller pixels than those of the previous camera generation emerged on the market. These cameras showed better performance and first applications in optical metrology became possible. Today, digital CCD cameras with 4 million pixels are standard.

The tremendous development in opto-electronics and in data processing pushed Digital Holography to new perspectives: It is applied with success in optical deformation and strain analysis, shape measurement, microscopy and for investigations of flows in liquids and gases. In this book we make the trial to describe the principles of this method and to report on the various applications. We took pains to prepare the manuscript carefully and to avoid mistakes. However, we are not perfect. Comments, suggestions for improvements or corrections are therefore welcome and will be considered in potential further editions.

Some pictures in this book originate from common publications with other co-authors. All of our co-workers, especially W. Osten, Th. Kreis, D. Holstein, S. Seebacher, H.-J. Hartmann and V. Kebbel are gratefully acknowledged.

Contents

1 Introduction... 1

2 Fundamental Principles of Holography 5
   2.1 Light Waves 5
   2.2 Interference... 8
   2.3 Coherence... 10
      2.3.1 General 10
      2.3.2 Temporal Coherence 11
      2.3.3 Spatial Coherence 13
   2.4 Diffraction 15
   2.5 Speckle 18
   2.6 Holography.. 20
      2.6.1 Hologram Recording and Reconstruction 20
      2.6.2 The Imaging Equations 23
   2.7 Holographic Interferometry 25
      2.7.1 Generation of Holographic Interferograms 25
      2.7.2 Displacement Measurement by HI................. 28
      2.7.3 Holographic Contouring....................... 30
      2.7.4 Refractive Index Measurement by HI............. 34
      2.7.5 Phase Shifting HI 35
      2.7.6 Phase Unwrapping.............................. 37

3 Digital Holography.................................... 39
   3.1 General Principles 39
   3.2 Numerical Reconstruction............................ 42
      3.2.1 Reconstruction by the Fresnel Approximation........ 42
      3.2.2 Reconstruction by the Convolution Approach 49
      3.2.3 Digital Fourier Holography...................... 52
   3.3 Shift and Suppression of DC-Term and Conjugate Image 53
      3.3.1 Suppression of the DC Term 53
      3.3.2 Tilted Reference Wave......................... 55
      3.3.3 Phase Shifting Digital Holography............... 56

3.4   Recording of Digital Holograms   58
      3.4.1   Image Sensors.   58
      3.4.2   Spatial Frequency Requirements   62
      3.4.3   Cameras for Digital Hologram Recording.   63
      3.4.4   Recording Set-ups.   64
      3.4.5   Stability Requirements.   66
      3.4.6   Light Sources   66

4   Digital Holographic Interferometry (DHI)   69
4.1   General Principles   69
4.2   Deformation Measurement   70
      4.2.1   Quantitative Displacement Measurement.   70
      4.2.2   Mechanical Materials Properties   74
      4.2.3   Thermal Materials Properties   78
      4.2.4   Non-destructive Testing   81
4.3   Shape Measurement.   85
      4.3.1   Two-Illumination-Point Method.   85
      4.3.2   Two- and Multi-wavelength Method   86
      4.3.3   Hierarchical Phase Unwrapping.   89
4.4   Measurement of Refractive Index Variations   92

5   Digital Holographic Particle Sizing and Microscopy.   95
5.1   Introduction   95
5.2   Recording and Replay Conditions   96
      5.2.1   In-line Recording   97
      5.2.2   Off-axis Recording   99
      5.2.3   Image Resolution   99
      5.2.4   Holographic Depth-of-Field and Depth-of-Focus 102
      5.2.5   Optical Aberrations 104
5.3   Data Processing and Autofocusing of Holographic Images 104
5.4   Some Applications in Imaging and Particle Sizing 105
      5.4.1   Particle Sizing. 106
      5.4.2   Digital Holographic Microscopy (DHM) 107
      5.4.3   Holographic Tomography. 110
      5.4.4   Phase Shifting DHM 113
      5.4.5   Particle Image Velocimetry (PIV) 114
      5.4.6   Underwater Digital Holography. 115

6   Special Techniques. 121
6.1   Applications Using Short Coherence Length Light 121
      6.1.1   Light-in-Flight Measurements 121
      6.1.2   Short-Coherence Tomography. 126
6.2   Endoscopic Digital Holography 127
6.3   Optical Reconstruction of Digital Holograms. 129

6.4   Comparative Digital Holography   131
        6.4.1   Fundamentals of Comparative Holography   131
        6.4.2   Comparative Digital Holography   132
  6.5   Encrypting of Information with Digital Holography   135
  6.6   Synthetic Aperture Holography   137
  6.7   Holographic Pinhole Camera   138

7   Computational Wavefield Sensing   141
  7.1   Overview   141
  7.2   Phase Retrieval   142
        7.2.1   Projection Based Methods   144
        7.2.2   Gradient Search Methods   154
        7.2.3   Deterministic Methods   156
  7.3   Shear Interferometry for Wavefield Sensing   162
        7.3.1   Wavefront Reconstruction   164
        7.3.2   Computational Shear Interferometry   177
  7.4   Shack-Hartmann Wavefront Sensing   183

8   Speckle Metrology   185
  8.1   Electronic Speckle Pattern Interferometry (ESPI)   185
  8.2   Digital Shearography   189
  8.3   Digital Speckle Photography   193
  8.4   Comparison of Conventional HI, ESPI and Digital HI   194

Appendix A: The Fourier Transform   199

Appendix B: Phase Transformation of a Spherical Lens   203

Appendix C: Simple Reconstruction Routines   207

References   211

Index   223

Chapter 1
Introduction

The recording and storage of full-parallax 3D images was and is a recurring goal of science and engineering since the first photographs were made. To accomplish this, the whole ("holos" in Greek) optical information emanating from a source needs to be written ("graphein" in Greek), recorded or captured on a sensing device for later recreation or reconstruction of the original object. This is the technique we now know as *holography*.

The history of holography started in principle when Lord Rayleigh experimentally created a Fresnel lens [268] and showed the generation of an interference pattern by the superposition of a spherical wave with a planar wave. In holography the planar wave is regarded as the reference wave and the spherical wave represents the object. The Fresnel lens in this sense can be regarded as the hologram of a point source. However, it was Denis Gabor who recognized that the same procedure carried out over a number of point's leads to the ability to optically reconstruct their position in space. Consequently, he coined the name "holography" since he was able to reconstruct the amplitude and phase of a wave [68–70].

A holographically stored image or hologram in the classical sense is a photographically, or otherwise, recorded interference pattern between a wave field scattered from an object and a coherent background denoted as the reference wave. A hologram is usually recorded on a flat two-dimensional surface, but contains the entire information about the three-dimensional wave field. This information is encoded in the form of interference fringes, usually not visible to the human eye due to their high spatial frequencies. The object wave can be recovered by illuminating the hologram with the original reference wave. This reconstructed wave is optically indistinguishable from the original object wave by passive means. An observer sees a three-dimensional image with all effects of perspective, parallax and depth-of-focus.

In his original set-up, Gabor illuminated the hologram with a parallel beam of light incident on a predominantly transparent object. The axes of both the object wave and the reference wave were parallel. The reconstruction of this hologram results in a real image superimposed on the undiffracted part of the reconstruction

© Springer-Verlag Berlin Heidelberg 2015
U. Schnars et al., *Digital Holography and Wavefront Sensing*,
DOI 10.1007/978-3-662-44693-5_1

wave and a so called 'twin image' (or virtual image) lying on the same optical axis, i.e. an in-line hologram. Significant improvements of this in-line geometry were proposed by Leith and Upatnieks [141, 142], who introduced an off-axis reference wave at an oblique angle; this wave does not pass through the object. This approach spatially separates the two images and the reconstruction wave and allows the capture of opaque objects.

One early application of classical holography is Holographic Interferometry (HI), developed in the late 1960s by Stetson, Powell [190, 222] and others. HI made it possible to map the displacements of rough surfaces with an accuracy of a fraction of a micrometer. It also enabled interferometric comparisons of stored wave fronts existing at different times.

The development of computer technology allowed transferring either the recording process or the reconstruction process into the computer. The first approach led to *Computer Generated Holography* (CGH), which artificially generates holograms by numerical methods followed by their optical reconstruction. This technique is not considered here and the interested reader is referred to the literature; see e.g. Lee [140], Bryngdahl and Wyrowski [20] or Schreier [204].

Numerical hologram reconstruction was initiated by Goodman and Lawrence [74] and Yaroslavskii et al. [132]. They sampled optically enlarged parts of in-line and Fourier holograms recorded on a photographic plate. These digitized conventional holograms were reconstructed numerically. Onural and Scott [146, 167, 168] improved the reconstruction algorithm and applied this method to particle measurement. Haddad et al. described a holographic microscope based on numerical reconstruction of Fourier holograms [78].

A big step forward in the 1990s was the development of direct recording of Fresnel holograms with Charged Coupled Devices (CCD's) by Schnars and Jüptner [197, 198]. This method enabled full digital recording and processing of holograms, without any photographic recording as intermediate step. The name which has been originally proposed for this technique was 'direct holography' [197], emphasizing the direct way from optical recording to numerical processing. Later on the term **Digital Holography** has been accepted in the optical metrology community for this method. Although this name is sometimes also used for Computer Generated Holography, the term Digital Holography is used within the scope of this book as a designation for digital recording and numerical reconstruction of holograms.

The dramatic developments in optics, electronics and computing widened the possibilities to capture, by computer means, phase information as well as amplitude. Computational wave front sensing [57] liberates the measurement procedures from a number of restrictions concerning coherence of the light or environmental requirements. It was shown that in some cases the light generated by the display of a smartphone has sufficient coherence to enable the recording of holograms [56]. Even the twin-image problem of in-line holography can be solved when phase shifted holograms are recorded according to Yamaguchi [255].

Schnars and Jüptner applied DH to interferometry and demonstrated that digital hologram reconstruction offers much more possibilities than conventional (optical) processing: The phase of the stored light waves can be calculated directly from

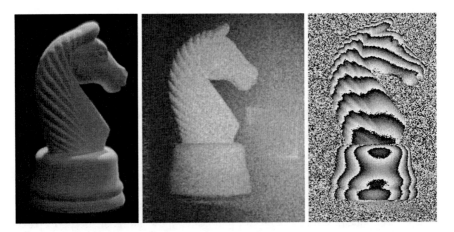

Fig. 1.1 Digital holography. *Left* photograph of a holographic reconstruction of a chess piece. *Middle* intensity reconstruction from a digital hologram. *Right* interference phase image after a thermal load is applied

digital holograms, without the need for generating phase shifted interferograms [195, 196], see example in Fig. 1.1. Other methods of optical metrology, such as shearography or speckle photography, can be derived numerically from digital holograms [199]. Using mathematical reconstruction, the choice of interferometric technique (hologram interferometry, shearography or other) can be left until after hologram recording.

The use of electronic devices such as CCDs to record interferograms was already established in Electronic Speckle Pattern Interferometry (*ESPI*, also named *TV-holography*). Proposed independently by Butters and Leendertz [23], Macovski et al. [150] and Schwomma [205], two speckle interferograms are recorded in different states of the object under investigation. The speckle patterns are subtracted electronically. The resulting fringe pattern has some similarities to that of conventional or digital HI. Digital Holographic Interferometry (DHI) and ESPI are competing methods: image subtraction in ESPI is easier than the numerical reconstruction of DHI, but the information content of digital holograms is higher. ESPI and other methods of speckle metrology are also discussed in this book in order to compare them with Digital Holographic Interferometry.

The main disadvantage of ESPI is the loss of phase information of the original wave in the correlation process [46, 147, 148]. The interference phase has to be recovered with phase shifting methods [35, 223, 224]. However, all the information can be reconstructed by evaluating phase shifted shearograms without the ESPI approach [57] leading to a wave sensing method with low demands on the coherence and the environmental requirements.

Since its inception Digital Holography has been extended, improved and applied to several measurement tasks. Some of these advances include:

- improvements of the experimental techniques and of the reconstruction algorithm [37, 39, 40, 75, 113, 122, 123, 126, 136, 182, 184, 203],
- applications in deformation analysis and shape measurement [34, 119, 171, 181, 200, 206, 246],
- the development of phase shifting digital holography [47, 103, 137, 255–258, 264],
- the development of Digital Holographic Microscopy [38, 43, 48, 94, 114, 183, 235–237, 253]
- applications in particle tracking and sizing and underwater holography [5, 6, 85, 127, 176, 214, 227],
- measurement of refractive index distributions within transparent media due to temperature or concentration variations [105, 106, 177, 252],
- applications in encrypting of information [95, 135, 231, 232],
- the development of digital light-in-flight holography and other short-coherence-length applications [25, 100, 162–164, 179, 189],
- the development of methods to reconstruct the three-dimensional object structure from digital holograms [63, 64, 96, 154, 233, 263]
- the development of comparative Digital Holography [173, 174]

A number of alternative concepts for wavefield sensing are based on computational methods. Here, in contrast to Digital Holography, determination of the complex amplitude of a wave field is treated as an inverse problem. The recorded intensities are interpreted as an effect which has been caused by an unknown wavefield that has undergone various manipulations. Examples include intensities corresponding to different propagation states or superposition of a wavefield with a shifted (or propagated) copy of itself. The great benefit is that no particular reference wave is required to measure the complex amplitude. In many situations, this makes computational methods not only more robust and flexible than Digital Holography but also enables application to wave fields with low spatial and/or temporal coherence. However, solving the inverse problem requires application of sophisticated numerical methods. In most cases there is no way to directly track back to the complex amplitude from the recorded intensities alone. It is also not possible to record on film material in order to optically reconstruct the investigated wave field. The evaluation procedure can therefore be regarded as an integral part of the measurement process.

As an introduction to the field, we will review the three methods of phase retrieval [71, 193, 260], shear interferometry [55–57] and Shack-Hartmann wavefield sensing in Chap. 7, which have been constantly developed since the early 1970s.

Chapter 2
Fundamental Principles of Holography

2.1 Light Waves

The behaviour of light can be modelled either as a propagating electromagnetic (e-m) wave or as a stream of massless particles known as photons. Although the models are seemingly contradictory both are necessary to fully describe the full gamut of light phenomena. Whichever model is most appropriate depends on the phenomenon to be described or the experiment under investigation. For example, interaction of light with the atomic structure of matter is best described by the photon model: the theory of photon behaviour and its interactions is known as quantum optics. The phenomenon of refraction, diffraction and interference, however, are best described in terms of the wave model i.e. classical electromagnetism.

Interference and diffraction form the basis of holography An e-m wave is described in terms of the propagation through space of mutually perpendicular electric and magnetic fields. These fields oscillate in a plane that is perpendicular to the direction of travel i.e. they are described as transverse waves, as depicted in Fig. 2.1. Light waves can be described either by the electrical or by the magnetic field, but in optics convention is to describe the e-m wave in terms of the electric vector.

Light propagation is described by the wave equation, which follows from Maxwell's equations. The wave equation in a vacuum is

$$\nabla^2 \vec{E} - \frac{1}{c^2} \frac{\partial^2 \vec{E}}{\partial t^2} = 0 \tag{2.1}$$

Here $\vec{E}$ is the electric field and ∇^2 is the *Laplace operator* defined as

$$\nabla^2 = \frac{\partial^2}{\partial x^2} + \frac{\partial^2}{\partial y^2} + \frac{\partial^2}{\partial z^2} \tag{2.2}$$

© Springer-Verlag Berlin Heidelberg 2015
U. Schnars et al., *Digital Holography and Wavefront Sensing*,
DOI 10.1007/978-3-662-44693-5_2

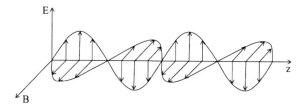

Fig. 2.1 Electromagnetic wave propagating in z-direction

and c is the speed of light in vacuum:

$$c = 2.9979 \times 10^8 \text{ m/s} \tag{2.3}$$

The electrical field $\vec{E}$ is a vector quantity and can vibrate in any direction perpendicular to the direction of propagation. However, in many applications the wave vibrates only in a single plane. Such light is called *linear polarized light*. In this case it is sufficient to consider the scalar wave equation

$$\nabla^2 E - \frac{1}{c^2}\frac{\partial^2 E}{\partial t^2} = 0 \tag{2.4}$$

It can be easily verified that a linearly polarized, harmonic plane wave with amplitude

$$E(x, y, z, t) = a \, \cos\left(\omega t - \vec{k}\vec{r} - \varphi_0\right) \tag{2.5}$$

is a solution of the above wave equation.

$E(x,y,z,t)$ is the modulus of the electrical field vector at the point with spatial vector $\vec{r} = (x, y, z)$ at the time t. The quantity a is the *amplitude* of the wave. The *wave vector* $\vec{k}$ describes the propagation direction of the wave:

$$\vec{k} = k\vec{n} \tag{2.6}$$

$\vec{n}$ is a unit vector in the propagation direction. Points of equal phase are located on parallel planes that are perpendicular to the propagation direction. The modulus of $\vec{k}$ is the *wave number* and is described by

$$\left|\vec{k}\right| \equiv k = \frac{2\pi}{\lambda} \tag{2.7}$$

The angular frequency ω corresponds to the frequency f of the light wave by

$$\omega = 2\pi f \tag{2.8}$$

Frequency f and wavelength λ are related through the speed of light c:

$$c = \lambda f \qquad (2.9)$$

The spatially varying term

$$\varphi = -\vec{k}\vec{r} - \varphi_0 \qquad (2.10)$$

is the *phase*, with phase constant φ_0. It has to be pointed out that this definition is not standardized. Some authors designate the entire argument of the cosine function, $\omega t - \vec{k}\vec{r} - \varphi_0$, as phase. The definition Eq. (2.10) is favourable to describe the holographic process and therefore used in this book.

The vacuum wavelengths of visible light are in the range of 400 nm (violet) to 780 nm (deep red). The corresponding frequency range is 7.5×10^{14} Hz to 3.8×10^{14} Hz. Light sensors such as the human eye, photodiodes, photographic film or CCD's are not able to detect such high frequencies due to technical and physical reasons. The only directly measurable quantity is the *intensity*. It is proportional to the time average of the square of the electrical field:

$$I = \varepsilon_0 c \langle E^2 \rangle_t = \varepsilon_0 c \lim_{T \to \infty} \frac{1}{2T} \int_{-T}^{T} E^2 dt \qquad (2.11)$$

$\langle E^2 \rangle_t$ denotes the time average over many light periods. The constant factor $\varepsilon_0 c$ results if the intensity is formally derived from the Maxwell equations. The constant ε_0 is the vacuum permittivity. Note: we are using the term intensity here. In photometry and radiometry *intensity* has a different meaning (radiant power per solid angle, unit W sr^{-1}).

For a plane wave Eq. (2.5) has to be inserted:

$$I = \varepsilon_0 c a^2 \left\langle \cos^2 \left(\omega t - \vec{k}\vec{r} - \varphi_0 \right) \right\rangle_t = \frac{1}{2} \varepsilon_0 c a^2 \qquad (2.12)$$

According to Eq. (2.12) the intensity is proportional to the square of the amplitude.

The expression (2.5) can be written in complex form as

$$E(x, y, z, t) = a \operatorname{Re}\left\{ \exp\left(i\left(\omega t - \vec{k}\vec{r} - \varphi_0 \right) \right) \right\} \qquad (2.13)$$

where 'Re' denotes the real part of the complex function. For computations the real part 'Re' can be omitted (in accordance with the superposition principle). However, only the real part represents the physical wave:

$$E(x,y,z,t) = a \exp\left(i\left(\omega t - \vec{k}\vec{r} - \varphi_0 \right) \right) \qquad (2.14)$$

One advantage of the complex representation is that the spatial and temporal parts factorize and Eq. (2.14) can be written as:

$$E(x,y,z,t) = a \, \exp(i\varphi) \exp(i\omega t) \qquad (2.15)$$

In many calculations of optics only the spatial distribution of the wave is of interest. In this case only the spatial part of the electrical field, its *complex amplitude*, need be considered:

$$A(x,y,z) = a \, \exp(i\varphi) \qquad (2.16)$$

Equations (2.15) and (2.16) are not just valid for plane waves, but apply in general to three-dimensional waves whose amplitude, a, and phase, φ, are functions of x, y and z.

In complex notation the intensity is now simply calculated by taking the square of the modulus of the complex amplitude

$$I = \frac{1}{2}\varepsilon_0 c|A|^2 = \frac{1}{2}\varepsilon_0 c A^* A = \frac{1}{2}\varepsilon_0 c a^2 \qquad (2.17)$$

where * denotes complex conjugation. In many practical calculations where the absolute value of I is not of interest the factor $\frac{1}{2}\varepsilon_0 c$ can be neglected, and the intensity simplifies to $I = |A|^2$.

2.2 Interference

The superposition of two or more waves in space is named *interference*. If each single wave described by $\vec{E}_i(\vec{r}, t)$ is a solution of the wave equation, the superposition

$$\vec{E}(\vec{r}, t) = \sum_i \vec{E}_i(\vec{r}, t) \quad i = 1, 2, \ldots \qquad (2.18)$$

is also a solution. This is because the wave equation is a linear differential equation.

In the following, interference of two monochromatic waves with equal frequencies and wavelengths is considered. The waves shall have the same polarization directions, i.e. scalar formalism can be used. The complex amplitudes of the respective waves are represented by;

$$A_1(x,y,z) = a_1 \exp(i\varphi_1) \qquad (2.19)$$

$$A_2(x,y,z) = a_2 \exp(i\varphi_2) \qquad (2.20)$$

The resulting complex amplitude is then calculated by the sum of the individual amplitudes:

$$A = A_1 + A_2 \tag{2.21}$$

According to Eq. (2.17) the intensity can be written as

$$
\begin{aligned}
I = |A_1 + A_2|^2 &= (A_1 + A_2)(A_1 + A_2)^* \\
&= a_1^2 + a_2^2 + 2a_1 a_2 \cos(\varphi_1 - \varphi_2) \\
&= I_1 + I_2 + 2\sqrt{I_1 I_2} \cos \Delta\varphi
\end{aligned}
\tag{2.22}
$$

where I_1, I_2 are the individual intensities and the phase difference between the two waves is

$$\Delta\varphi = \varphi_1 - \varphi_2 \tag{2.23}$$

The resulting intensity is the sum of the individual intensities *plus* the inter-ference term $2\sqrt{I_1 I_2} \cos \Delta\varphi$, which depends on the phase difference between the waves. The intensity reaches its maximum when the phase difference between consecutive points is a multiple of 2π

$$\Delta\varphi = 2n\pi \quad \text{for } n = 0, 1, 2, \ldots \tag{2.24}$$

This is known as *constructive interference*. The intensity reaches its minimum when

$$\Delta\varphi = (2n + 1)\pi \quad \text{for } n = 0, 1, 2, \ldots \tag{2.25}$$

And this is known as *destructive interference*. The integer n is the interference order. An interference pattern therefore consists of a series of dark and light lines, "fringes", across the field-of-view as a result of this constructive and destructive interference. Scalar theory can also be applied to waves with different polarization directions, if the components of the electric field vector are considered.

The superposition of two plane waves which intersect at an angle θ with respect to each other results in an interference pattern with equidistant spacing, as seen in Fig. 2.2. The fringe spacing d is the distance from one interference maximum to the next and can be calculated from geometrical considerations. Figure 2.2 shows that

$$\sin \theta_1 = \frac{\Delta l_1}{d}; \quad \sin \theta_2 = \frac{\Delta l_2}{d} \tag{2.26}$$

The quantities θ_1 and θ_2 are the angles between the propagation directions of the wavefronts and the vertical direction of the screen. The length Δl_2 is the path difference between wavefront W2 and wavefront W1 at the position of the inter-ference maximum P1 (W2 has to travel a longer path to P1 than W1). At the neighboring maximum P2 the conditions are exchanged: now W1 has to travel a longer path; the path difference of W2 with respect to W1 is $-\Delta l_1$. The variation

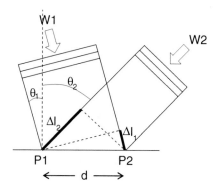

Fig. 2.2 Interference of two plane waves W1 and W2. The marginal rays are sketched. θ_1 is the angle between W1 and the vertical, θ_2 is the angle between W2 and the vertical. P1 and P2 are adjacent interference maxima

between the path differences at neighboring maxima is therefore $\Delta l_1 + \Delta l_2$. This difference is equal to one wavelength. Thus the interference condition is:

$$\Delta l_1 + \Delta l_2 = \lambda \tag{2.27}$$

Combining Eq. (2.26) with Eq. (2.27) gives the fringe spacing as:

$$d = \frac{\lambda}{\sin\theta_1 + \sin\theta_2} = \frac{\lambda}{2\sin\frac{\theta_1+\theta_2}{2}\cos\frac{\theta_1-\theta_2}{2}} \tag{2.28}$$

The approximation $\cos(\theta_1 - \theta_2)/2 \approx 1$ and $\theta = \theta_1 + \theta_2$ can be applied to give

$$d = \frac{\lambda}{2\sin\frac{\theta}{2}} \tag{2.29}$$

Instead of the fringe spacing d, the fringe pattern can also be described in terms of the spatial frequency f, which is just the reciprocal of d, i.e.

$$f = d^{-1} = \frac{2}{\lambda}\sin\frac{\theta}{2} \tag{2.30}$$

2.3 Coherence

2.3.1 General

Generally the resulting intensity of two different sources, e.g. two electric light bulbs directed on a screen, is additive. Instead of dark and bright fringes as expected by Eq. (2.22) only a uniform brightness according to the sum of the individual intensities is visible.

In order to observe interference fringes, the phases of the individual waves have to be correlated. The ability of light to form interference patterns is called *coherence* and is investigated in this chapter. The two aspects of coherence are temporal and spatial coherence. Temporal coherence depends on the correlation of a wave with itself at different instants in time [121], whereas spatial coherence is based on the mutual correlation of different parts of the same wavefield in space.

2.3.2 Temporal Coherence

The phenomenon of interference between two coherent beams of light can be described in terms of a two beam interferometer such as the Michelson-interferometer, as shown in Fig. 2.3. Light emitted by the source S is split into two waves of reduced amplitude by the beam splitter BS. These waves travel to the mirrors M1 and M2 respectively, and are reflected back into their incident directions. After passing the beam splitter again they are superimposed at a screen. Usually the superimposed waves are not exactly parallel, but are incident at a small angle. As a result a two-dimensional interference pattern becomes visible.

The optical path length from BS to M1 and back to BS is s_1, and the optical path length from BS to M2 and back to BS is s_2. Experiments show that interference can only occur if the optical path difference $s_1 - s_2$ does not exceed a certain length L. If the optical path difference exceeds this limit, the interference fringes vanish and just a uniform brightness is visible on the screen. The qualitative explanation for this phenomenon is that interference fringes can only develop if the superimposed waves have a well defined (constant) phase relationship between them. The phase difference between waves emitted by different sources varies randomly and thus the waves do not interfere. The atoms within the light source emit wave trains with a finite length L. If the optical path difference exceeds L, the recombined waves do not overlap after passing the different ways and interference is not observed.

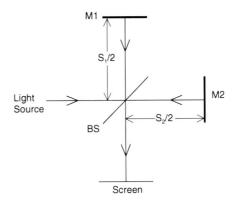

Fig. 2.3 Michelson's interferometer

The critical path length difference or, equivalently, the length of a wave train is the *coherence length L* of the wave. The corresponding time over which the wave train is emitted is its *coherence time,*

$$\tau = \frac{L}{c} \qquad (2.31)$$

According to the laws of Fourier analysis a wave train with finite length L corresponds to light with finite spectral width Δf, where

$$L = \frac{c}{\Delta f} \qquad (2.32)$$

The coherence length is therefore a measure for the spectral linewidth of the source at a specific frequency, f. Light with a long coherence length accordingly has a correspondingly small linewidth and is therefore highly monochromatic.

Typical coherence lengths of light radiated from thermal sources, e.g. conventional electric light bulbs, are in the range of some micrometers. That means, interference can only be observed if the arms of the interferometer have nearly equal path lengths. On the other hand lasers have coherence lengths from a few millimetres (e.g. a multi-mode diode laser) to several 100 m (e.g. a stabilized single mode Nd:YAG-laser) up to several hundred kilometres for specially stabilized gas lasers used for research purposes.

The fringe visibility

$$V = \frac{I_{max} - I_{min}}{I_{max} + I_{min}} \qquad (2.33)$$

is a measure of the contrast of a particular interference pattern, where I_{max} and I_{min} are two neighbouring intensity maxima and minima. They are calculated by inserting $\Delta\varphi = 0$ and $\Delta\varphi = \pi$ respectively into Eq. (2.22). In the ideal case of infinite coherence length the visibility is given by,

$$V = \frac{2\sqrt{I_1 I_2}}{I_1 + I_2} \qquad (2.34)$$

To consider the effect of finite coherence length the *complex self-coherence function* $\Gamma(\tau)$ is introduced:

$$\Gamma(\tau) = \langle E(t + \tau)E^*(t) \rangle$$

$$= \lim_{T \to \infty} \frac{1}{2T} \int_{-T}^{T} E(t + \tau)E^*(t)dt \qquad (2.35)$$

$E(t)$ is the electrical field (to be precise: the complex analytical signal) of one interfering wave while $E(t + \tau)$ is the electrical field of the other wave. The latter is delayed in time by τ. Equation (2.35) represents the autocorrelation of the corresponding electric field amplitudes. The quantity

$$\gamma(\tau) = \frac{\Gamma(\tau)}{\Gamma(0)} \tag{2.36}$$

is the normalized self-coherence function; the absolute value of γ defines the degree of coherence.

With finite coherence length the interference equation (2.22) has to be replaced by

$$I = I_1 + I_2 + 2\sqrt{I_1 I_2}|\gamma| \cos \Delta\varphi \tag{2.37}$$

The maximum and minimum intensity are now calculated by

$$\begin{aligned} I_{\max} &= I_1 + I_2 + 2\sqrt{I_1 I_2}|\gamma| \\ I_{\min} &= I_1 + I_2 - 2\sqrt{I_1 I_2}|\gamma| \end{aligned} \tag{2.38}$$

Inserting these quantities into Eq. (2.33) yields

$$V = \frac{2\sqrt{I_1 I_2}}{I_1 + I_2}|\gamma| \tag{2.39}$$

For two partial waves with the same intensity, $I_1 = I_2$ Eq. (2.39) becomes

$$V = |\gamma| \tag{2.40}$$

$|\gamma|$ is equal to the visibility and is therefore a measure of the ability of the two wave fields to interfere. When $|\gamma| = 1$ we have ideally monochromatic light or, likewise, light with infinite coherence length; when $|\gamma| = 0$ for the light is completely incoherent. Partially coherent light therefore lies in the range $0 < |\gamma| < 1$.

2.3.3 Spatial Coherence

Spatial coherence describes the mutual correlation of spatially separated parts of the same wavefield. This property can be measured using, for example, a Young's interferometer, Fig. 2.4. Here, an extended light source emits light from a large number of elementary point sources. An aperture with two transparent holes is mounted between the light source and the screen. The aim of the experiment is to determine the mutual correlation (degree of coherence) of the light incident on the aperture at the spatially separated positions given by the holes. If the light at these

Fig. 2.4 Young's
interferometer

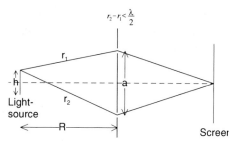

positions is correlated, interference fringes are visible on the screen. In the following we will discuss the geometrical relations under which interference can be observed for the simple case of an extended light source. The fringes result from the different light paths traversed to the screen, either via the upper or via the lower hole in the aperture [250]. The interference pattern vanishes if the distance between the holes a exceeds the critical limit a_k. This limit is named *coherence distance*. The phenomenon is not related to the spectral width of the light source, but is due to the waves emitted by different points of the extended light source being superimposed on the screen. It may happen that a particular source point generates an interference maximum at a certain point on the screen, while another source point generates a minimum at the same point. This occurs because the optical path difference is different for each source point. In general the contributions from all source points cancel and the contrast vanishes. This cancellation is avoided if the following condition is fulfilled for every point of the light source:

$$r_2 - r_1 < \frac{\lambda}{2} \tag{2.41}$$

This condition is fulfilled if it is restricted to rays emanating from the edges of the light source. The following relations are valid for points at the edges:

$$r_1^2 = R^2 + \left(\frac{a-h}{2}\right)^2; \quad r_2^2 = R^2 + \left(\frac{a+h}{2}\right)^2 \tag{2.42}$$

where h is the width of the light source. Applying the assumptions $a \ll R$ and $h \ll R$ gives,

$$r_2 - r_1 \approx \frac{ah}{2R} \tag{2.43}$$

Combining Eqs. (2.41) and (2.43) leads to the following expression:

$$\frac{ah}{2R} < \frac{\lambda}{2} \tag{2.44}$$

The coherence distance a_k is therefore given from,

$$\frac{a_k h}{2R} = \frac{\lambda}{2} \tag{2.45}$$

In contrast to temporal coherence, the spatial coherence depends not only on properties of the light source, but also on the geometry of the interferometer. A light source may initially generate interference, which means Eq. (2.44) is fulfilled, but if the distance between the holes increases or the distance between the light source and the aperture decreases, Eq. (2.44) is violated and the interference vanishes.

To consider spatial coherence the autocorrelation function defined in Eq. (2.35) is extended to,

$$\Gamma(\vec{r}_1, \vec{r}_2, \tau) = \langle E(\vec{r}_1, t + \tau) E^*(\vec{r}_2, t) \rangle$$

$$= \lim_{T \to \infty} \frac{1}{2T} \int_{-T}^{T} E(\vec{r}_1, t + \tau) E^*(\vec{r}_2, t) dt \tag{2.46}$$

where $\vec{r}_1$, $\vec{r}_2$ are the spatial vectors of the holes in the aperture of the Young interferometer. This cross correlation function is the *mutual coherence function*. The normalized function is

$$\gamma(\vec{r}_1, \vec{r}_2, \tau) = \frac{\Gamma(\vec{r}_1, \vec{r}_2, \tau)}{\sqrt{\Gamma(\vec{r}_1, \vec{r}_1, 0) \Gamma(\vec{r}_2, \vec{r}_2, 0)}} \tag{2.47}$$

where $\Gamma(\vec{r}_1, \vec{r}_1, 0)$ is the intensity at $\vec{r}_1$ and $\Gamma(\vec{r}_2, \vec{r}_2, 0)$ is the intensity at $\vec{r}_2$. Equation (2.47) describes the degree of correlation between the lightfield at $\vec{r}_1$ at a time $t + \tau$ with the light field at $\vec{r}_2$ at time t. The special function $\gamma(\vec{r}_1, \vec{r}_2, \tau = 0)$ is a measure for the correlation between the field amplitudes at $\vec{r}_1$ and $\vec{r}_2$ at the same time and is defined as the *complex degree of coherence*. The modulus of the mutual coherence function $|\gamma(\vec{r}_1, \vec{r}_2, \tau)|$ is measured with the Young interferometer.

2.4 Diffraction

Consider a light wave incident on an obstacle such as an opaque screen with some holes, or *vice versa*, a transparent medium with opaque obstructions. From geometrical optics it is known that a shadow is visible on a screen behind the obstacle. On closer examination, we see that if the dimensions of the obstacle (e.g. diameter of holes in an opaque screen or size of opaque particles in a transparent volume) are of the order of the wavelength of the incident light, then the light distribution is not sharply bounded, but forms a pattern of dark and bright regions. This is the phenomenon *diffraction*, see Fig. 2.5.

Fig. 2.5 Diffraction of a
plane wave at an opaque
screen with a small hole

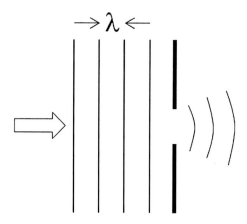

Diffraction can be explained qualitatively with the *Huygens' principle*: *Every
point of a wave front can be considered as a source point for secondary spherical
waves. The wave field at any other place is the coherent superposition of these
secondary waves.*

Huygens' principle is illustrated in Fig. 2.6.

The Fresnel-Kirchhoff integral describes diffraction quantitatively [116] as,

$$\Gamma(\xi', \eta') = \frac{i}{\lambda} \int\limits_{-\infty}^{\infty} \int\limits_{-\infty}^{\infty} A(x,y) \frac{\exp\left(-i\frac{2\pi}{\lambda}\rho'\right)}{\rho'} Q \, dx \, dy \qquad (2.48)$$

Fig. 2.6 Huygens' principle

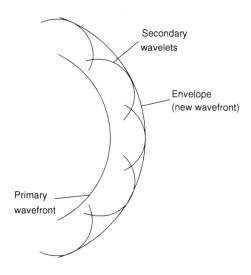

with

$$\rho' = \sqrt{(x - \xi')^2 + (y - \eta')^2 + d^2} \qquad (2.49)$$

and

$$Q = \frac{1}{2}(\cos\theta + \cos\theta') \qquad (2.50)$$

$A(x, y)$ is the complex amplitude in the plane of the diffracting aperture, see the coordinate system defined in Fig. 2.7. $\Gamma(\xi', \eta')$ is the complex amplitude in the observation plane. The term ρ' is the distance between a point in the aperture plane and a point in the observation plane.

Equation (2.48) can be understood as the mathematical formulation of Huygens' principle. The light source S lying in the source plane with coordinates (ξ, η) radiates spherical waves. $A(x,y)$ is the complex amplitude of such a wave in the aperture plane. At first an opaque aperture with only one hole at the position (x,y) is considered. Such a hole is now the source for secondary waves. The field at the position (ξ', η') of the diffraction plane is proportional to the field at the entrance side of the aperture $A(x,y)$ and to the field of the secondary spherical wave emerging from (x,y), described by $\exp(-i2\pi/\lambda\rho')/\rho'$. Now the entire aperture as a plane consisting of many sources for secondary waves is considered. The entire resulting field in the diffraction plane is therefore the integral over all secondary spherical waves, emerging from the aperture plane.

From the Huygens' principle it follows that the secondary waves not only propagate in the forward direction, but also back towards the source. Yet, experiment demonstrates that the wavefronts always propagate in one direction. To exclude this unrealistic situation the inclination factor Q defined in Eq. (2.50) is formally introduced into the Fresnel-Kirchhoff integral. Q depends on the angle θ between the incident light from the source and the unit vector $\vec{n}$ perpendicular to the aperture plane, and on the angle θ' between the diffracted light and $\vec{n}$, see Fig. 2.8. Q is approximately zero for $\theta \approx 0$ and $\theta' \approx \pi$. This excludes the concept of waves travelling in the backward direction. In most practical situations both θ and θ' are

Fig. 2.7 Coordinate system

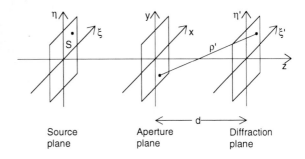

Source plane   Aperture plane   Diffraction plane

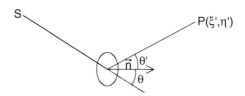

Fig. 2.8 Propagation geometry

very small and $Q \approx 1$. The inclination factor can be considered as an *ad hoc* correction to the diffraction integral, as done here, or be derived in the formal diffraction theory [73, 116].

Some authors use a "+" sign in the argument of the exponential function of the Fresnel-Kirchhoff integral $[\Gamma(\xi, \eta) = \ldots A(x, y) \exp(+i2\pi/\lambda\rho')/\rho' \ldots]$ instead of the "−" sign used here. This is dependent on whether we define the harmonic wave in Eq. (2.14), as either $\exp(+i\varphi)$ or $\exp(-i\varphi)$. However, using the "+" sign in Eq. (2.48) leads to the same expressions for all measurable quantities, as e.g. the intensity and the magnitude of the interference phase used in Digital Holographic Interferometry.

2.5 Speckle

A rough surface illuminated with coherent light always appears "grainy" or "speckly" to an observer. This is due to the random fluctuations in intensity of the light scattered from the surface and gives rise to a series of and dark and bright spots or known as *speckle*, and forms a speckle pattern across the surface (Fig. 2.9). A speckle pattern develops if the height variations of the rough surface are larger than the wavelength of the light.

Speckle results from interference of light scattered by the surface points. The phase of the waves scattered by different surface points fluctuate statistically due to the height variations. If these waves interfere with each other, a stationary speckle pattern is observed.

Fig. 2.9 A speckle pattern from a rough surface under coherent illumination

It can be shown that the probability density function for the intensity in a speckle pattern obeys negative exponential statistics [72]:

$$P(I)dI = \frac{1}{\langle I \rangle} \exp\left(-\frac{I}{\langle I \rangle}\right) \qquad (2.51)$$

$P(I)dI$ is the probability that the intensity at a certain point is lies between I and $I + dI$. $\langle I \rangle$ is the mean intensity of the entire speckle field. The most probable intensity value of a speckle is therefore zero, i.e. most speckles are black. The standard deviation σ_I is calculated by

$$\sigma_I = \langle I \rangle \qquad (2.52)$$

That means the intensity variations are in the same order as the mean value. The usual definition of the contrast of the speckle pattern is

$$V = \frac{\sigma_I}{\langle I \rangle} \qquad (2.53)$$

The contrast of a speckle pattern is therefore always unity.

One can distinguish between *objective* and *subjective* speckle formation. An objective speckle pattern develops on a screen, located in a distance z from the illuminated surface, Fig. 2.10. There is no imaging system between the surface and the screen. The size of an individual speckle in an objective speckle pattern can be estimated using the spatial frequency formula of Eq. (2.30). The two edge points of the illuminated surface form the highest spatial frequency given as,

$$f_{max} = \frac{2}{\lambda} \sin\frac{\theta_{max}}{2} \approx \frac{L}{\lambda z} \qquad (2.54)$$

The reciprocal of f_{max} is a measure for the speckle size; and hence the diameter of the speckle is,

Fig. 2.10 Objective speckle formation

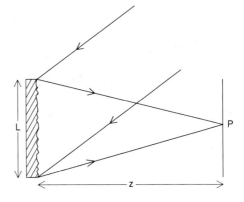

Fig. 2.11 Subjective speckle
formation

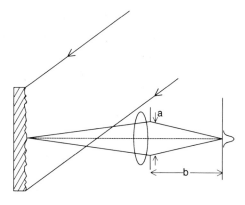

$$d_{Sp} = \frac{\lambda z}{L} \tag{2.55}$$

A subjective speckle pattern develops if the illuminated surface is focused with an imaging system, e.g. a camera lens or the human eye, as in Fig. 2.11. In this case the speckle diameter depends on the aperture diameter a of the imaging system. The size of a speckle in a subjective speckle pattern can be estimated again using the spatial frequency:

$$f_{max} = \frac{2}{\lambda} \sin\left(\frac{\theta_{max}}{2}\right) \approx \frac{a}{\lambda b} \tag{2.56}$$

where b is the image distance of the imaging system. It follows that the speckle diameter is given by

$$d_{Sp} = \frac{\lambda b}{a} \tag{2.57}$$

The speckle size can be increased by reducing the aperture of the imaging system.

2.6 Holography

2.6.1 Hologram Recording and Reconstruction

Holograms are usually recorded with an optical set-up consisting of a light source (e.g. a laser), mirrors and lenses for beam guiding and a recording device (e.g. a photographic sensor). A typical set-up is shown in Fig. 2.12 [79, 121]. Light with sufficient coherence is split into two waves of reduced amplitude by a beam splitter

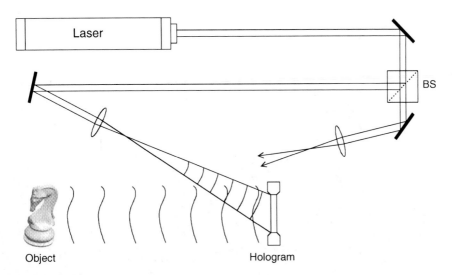

Object   Hologram

Fig. 2.12 Hologram recording

(BS). The first wave illuminates the object, is scattered at the object surface and reflected towards the recording medium. The second wave—the reference wave—directly illuminates the light sensitive medium. The waves interfere with each other to produce a characteristic interference pattern. In classical photographic holography the interference pattern is recorded on a photosensitive material such as silver halide films or plates and rendered permanent by wet chemical development of the film. In digital holography the interference pattern is recorded directly onto an electronic photosensor such as a CCD or CMOS array. The recorded interference pattern is the hologram.

The original object wave is reconstructed by illuminating the hologram with the reference wave, Fig. 2.13. An observer sees a virtual image, which is optically indistinguishable from the original object. The reconstructed image exhibits all effects of perspective, parallax and depth-of-field.

The holographic process is described mathematically using the formalism of Sect. 2.2. Across the extent of the photographic plate, the complex amplitude of the object wave is described by

$$E_O(x,y) = a_O(x,y) \exp(i\varphi_O(x,y)) \tag{2.58}$$

with real amplitude a_O and phase φ_O.

$$E_R(x,y) = a_R(x,y) \exp(i\varphi_R(x,y)) \tag{2.59}$$

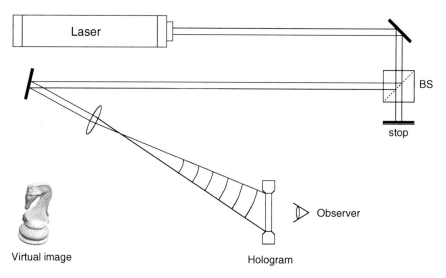

Fig. 2.13 Hologram reconstruction

is the complex amplitude of the reference wave with real amplitude a_R and phase φ_R.

Both waves interfere at the surface of the recording medium and the resultant intensity is described by

$$
\begin{aligned}
I(x,y) &= |E_O(x,y) + E_R(x,y)|^2 \\
&= (E_O(x,y) + E_R(x,y))(E_O(x,y) + E_R(x,y))^* \\
&= E_R(x,y)E_R^*(x,y) + E_O(x,y)E_O^*(x,y) + E_O(x,y)E_R^*(x,y) + E_R(x,y)E_O^*(x,y)
\end{aligned}
$$

$$(2.60)$$

The amplitude transmission $h(x, y)$ of the developed photographic plate (or of other recording media) is proportional to $I(x, y)$:

$$
h(x,y) = h_0 + \beta\tau I(x,y) \tag{2.61}
$$

The constant β is the slope of the amplitude transmittance versus exposure characteristic of the light sensitive material. For photographic emulsions β is negative. The exposure duration is denoted by τ and h_0 is the amplitude transmission of the unexposed plate; $h(x,y)$ is the hologram function. In Digital Holography using CCD or CMOS arrays as the recording medium, h_0 can be neglected.

For hologram reconstruction in classical holography, the hologram is illuminated with a replica of the original reference wave in terms of wavelength and phase. This is represented mathematically as a multiplication of the amplitude transmission of the medium with the complex amplitude of the reconstruction (reference) wave,

$$E_R(x,y)h(x,y) = \left[h_0 + \beta\tau\left(a_R^2 + a_O^2\right)\right]E_R(x,y)$$
$$+ \beta\tau a_R^2 E_O(x,y) + \beta\tau E_R^2(x,y)E_O^*(x,y) \qquad (2.62)$$

The first term on the right side of this equation is the reference wave multiplied by a constant factor. It represents the non-diffracted wave passing through the hologram (zero diffraction order). The second term is the reconstructed object wave and forms the virtual image. The real factor $\beta\tau a_R^2$ only influences the brightness of the image. The third term generates a distorted real image of the object. For off-axis holography the virtual image, the real image and the non-diffracted wave are spatially separated.

The reason for the distortion of the real image is the spatially varying complex factor E_R^2, which modulates the image forming conjugate object wave E_O^*. An undistorted real image can be generated by replaying the hologram with the complex conjugate of the reference beam E_R^*. This is mathematically represented by,

$$E_R^*(x, y)h(x, y) = \left[h_0 + \beta\tau\left(a_R^2 + a_O^2\right)\right]E_R^*(x, y)$$
$$+ \beta\tau a_R^2 E_O^*(x, y) + \beta\tau E_R^{*2}(x, y)E_O(x, y) \qquad (2.63)$$

2.6.2 The Imaging Equations

The virtual image appears at the position of the original object if the hologram is reconstructed with the same parameters as those used in the recording process. However, if one changes the wavelength or the coordinates of the reconstruction wave source point with respect to the coordinates of the reference wave source point used in the recording process, the position of the reconstructed image moves. The coordinate shift is different for all points, thus the shape of the reconstructed object is distorted. The image magnification is also influenced by the reconstruction parameters.

The *imaging equations* relate the coordinates of an object point O to those of the corresponding point in the reconstructed image. These equations are quoted here without derivation but are described in some detail in other textbooks [79, 121].

The coordinate system is shown in Fig. 2.14. The coordinates of the object point O are denoted as (x_O, y_O, z_O), (x_R, y_R, z_R) are the coordinates of the source point of the reference wave used for hologram recording and (x_P, y_P, z_P) are the coordinates of the source point of the reconstruction wave. The ratio between the recording wavelength λ_1 and the reconstruction wavelength λ_2 is denoted by $\mu = \lambda_2/\lambda_1$. The coordinates of the point in the reconstructed virtual image, which corresponds to the object point O, are:

$$x_1 = \frac{x_P z_O z_R + \mu x_O z_P z_R - \mu x_R z_P z_O}{z_O z_R + \mu z_P z_R - \mu z_P z_O} \qquad (2.64)$$

Fig. 2.14 Coordinate system used to describe holographic reconstruction. **a** Hologram recording. **b** Image reconstruction

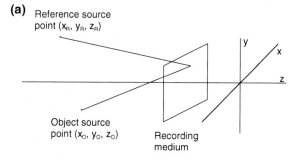

(a) Reference source point (x_R, y_R, z_R)

Object source point (x_O, y_O, z_O) Recording medium

y x z

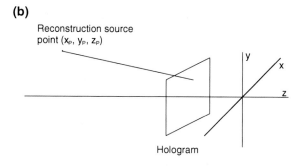

(b) Reconstruction source point (x_P, y_P, z_P)

Hologram

y x z

$$y_1 = \frac{y_P z_O z_R + \mu y_O z_P z_R - \mu y_R z_P z_O}{z_O z_R + \mu z_P z_R - \mu z_P z_O} \tag{2.65}$$

$$z_1 = \frac{z_P z_O z_R}{z_O z_R + \mu z_P z_R - \mu z_P z_O} \tag{2.66}$$

The coordinates of the point in the reconstructed real image, which corresponds to the object point O, are:

$$x_2 = \frac{x_P z_O z_R - \mu x_O z_P z_R + \mu x_R z_P z_O}{z_O z_R - \mu z_P z_R + \mu z_P z_O} \tag{2.67}$$

$$y_2 = \frac{y_P z_O z_R - \mu y_O z_P z_R + \mu y_R z_P z_O}{z_O z_R - \mu z_P z_R + \mu z_P z_O} \tag{2.68}$$

$$z_2 = \frac{z_P z_O z_R}{z_O z_R - \mu z_P z_R + \mu z_P z_O} \tag{2.69}$$

An extended object can be considered to be made up of a number of point objects. The coordinates of all surface points are described by the above equations. The lateral magnification of the entire virtual image is:

$$M_{lat,1} = \frac{dx_1}{dx_O} = \left[1 + z_0\left(\frac{1}{\mu z_P} - \frac{1}{z_R}\right)\right]^{-1} \tag{2.70}$$

The lateral magnification of the real image is given by,

$$M_{lat,2} = \frac{dx_2}{dx_O} = \left[1 - z_0\left(\frac{1}{\mu z_P} + \frac{1}{z_R}\right)\right]^{-1} \tag{2.71}$$

The longitudinal magnification of the virtual image is given by:

$$M_{long,1} = \frac{dz_1}{dz_O} = \frac{1}{\mu}M_{lat,1}^2 \tag{2.72}$$

The longitudinal magnification of the real image is:

$$M_{long,2} = \frac{dz_2}{dz_O} = -\frac{1}{\mu}M_{lat,2}^2 \tag{2.73}$$

There is a difference between real and virtual image which should be noted: since the real image is formed by the conjugate object wave O^*, it has the curious property that its depth is inverted. Corresponding points of the virtual image (which coincide with the original object points) and of the real image are located at equal distances from the hologram plane, but at opposite sides of it. The background and the foreground of the real image are therefore exchanged. The real image appears with the "wrong perspective". It is called a *pseudoscopic image*, in contrast to a normal or *orthoscopic* image.

2.7 Holographic Interferometry

2.7.1 Generation of Holographic Interferograms

Holographic Interferometry (HI) is a method of measuring optical path length variations, which are caused by deformations of opaque bodies or refractive index variations in transparent media, e.g. fluids or gases [175]. HI is a non-contact, non-destructive metrological technique with a very high measurement sensitivity. Optical path changes up to one hundredth of a wavelength are resolvable.

Two coherent wave fields, which are reflected from an object when it is in two different states of excitation, interfere. This is achieved e.g. in double-exposure holography by the recording of two wave fields on a single photographic plate, Fig. 2.15. The first exposure represents the object in its reference state (undeformed state), the second exposure represents the object in its loaded (deformed) state. The hologram is reconstructed by illumination with the reference wave, Fig. 2.16. As a result of the superposition of two holographic recordings with *slightly* different

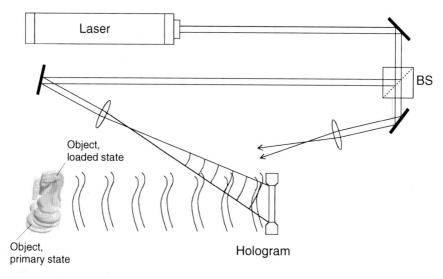

Fig. 2.15 Recording of a double exposed hologram

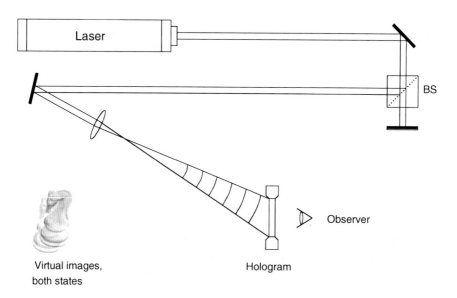

Fig. 2.16 Reconstruction of the double-exposed hologram

Fig. 2.17 A holographic interferogram of a pressure vessel

object waves, only one image superimposed by interference fringes is visible, see example in Fig. 2.17. From this holographic interferogram the observer can determine optical path changes due to the object deformation or other effects.

In real time HI, the hologram is replaced—after chemical processing—in exactly the original recording position. When it is illuminated with the reference wave, the reconstructed virtual image coincides with the object and is superimposed upon it. Interference patterns caused by phase changes between the holographically reconstructed reference object wave and the actual object wave are observable in real time.

The following mathematical description is valid for both the double exposure and real time techniques. The complex amplitude of the object wave in its initial state is:

$$E_1(x,y) = a(x,y)\exp[i\varphi(x,y)] \qquad (2.74)$$

where $a(x,y)$ is the real amplitude and $\varphi(x,y)$ is the phase of the object wave.

Optical path changes due to deformations of the object surface can be described by a variation of the phase from φ to $\varphi + \Delta\varphi$. The term $\Delta\varphi$ represents the difference between the reference and the actual phase and is known as the *interference phase*. The complex amplitude of the actual object wave is therefore denoted by

$$E_2(x,y) = a(x,y)\exp[i(\varphi(x,y) + \Delta\varphi(x,y))] \qquad (2.75)$$

The intensity of a holographic interference pattern is described by the square of the sum of the complex amplitudes. It is calculated as follows:

$$\begin{aligned} I(x,y) &= |E_1 + E_2|^2 = (E_1 + E_2)(E_1 + E_2)^* \\ &= 2a^2(1 + \cos(\Delta\varphi)) \end{aligned} \qquad (2.76)$$

The general expression for the intensity within an interference pattern is therefore:

$$I(x,y) = A(x,y) + B(x,y) \cos \Delta\varphi(x,y) \qquad (2.77)$$

The parameters $A(x,y)$ and $B(x,y)$ depend on the coordinates in the interferogram.

In practice these parameters are not known due to several disturbing effects, such as,

- uneven illumination of the object due to the Gaussian profile the expanded laser beam gives rise to varying brightness of the holographic interferogram.
- high frequency speckle noise is superimposed upon interferogram.
- additional superimposed diffraction patterns due to dust particles in the optical path.
- the varying reflectivity of the object under investigation may influence the brightness and visibility of the interferogram.
- electronic recording and transmission of holographic interferograms can generate additional noise.

Equation (2.77) describes the relation between the intensity of the interference pattern and the interference phase, which contains the information about the physical quantity to be measured (object displacement, refractive index change or object shape). In general it is not possible to calculate $\Delta\varphi$ directly from the measured intensity, because the parameters $A(x, y)$ and $B(x, y)$ are not known. In addition the cosine is an even function ($\cos 30° = \cos -30°$) and the sign of $\Delta\varphi$ cannot be determined unambiguously. Therefore several techniques have been developed to determine the interference phase by recording additional information. The most common techniques are the various phase shifting methods, which are briefly discussed in Sect. 2.7.5.

2.7.2 Displacement Measurement by HI

In this chapter a relationship between the measured interference phase and the displacement of the object surface under investigation is derived [121, 218]. The geometric quantities are explained in Fig. 2.18. The vector $\vec{d}(x,y,z)$ is the displacement vector. It describes the shift of a surface point from its initial position P_1 to the new position P_2 due to deformation. The terms $\vec{s}_1$ and $\vec{s}_2$ are unit vectors from the illumination source point S to P_1, and P_2 respectively. Similarly, $\vec{b}_1$ and $\vec{b}_2$ are unit vectors from P_1 to the observation point B, and from P_2 to B, respectively. The optical path difference between a ray from S to B via P_1 and a ray from S to B via P_2 is therefore given by,

$$\begin{aligned}
\delta &= \overline{SP_1} + \overline{P_1B} - \left(\overline{SP_2} + \overline{P_2B}\right) \\
&= \vec{s_1}\,\overrightarrow{SP_1} + \vec{b_1}\,\overrightarrow{P_1B} - \vec{s_2}\,\overrightarrow{SP_2} - \vec{b_2}\,\overrightarrow{P_2B}
\end{aligned} \qquad (2.78)$$

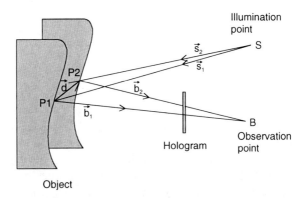

Fig. 2.18 Calculation of the interference phase

The lengths $\overline{SP_{1/2}}$ and $\overline{P_{1/2}B}$ are in the range of metres, while $\left|\vec{d}\right|$ is in the range of several micrometres. The vectors $\vec{s}_1$ and $\vec{s}_2$ can therefore be replaced by a unit vector $\vec{s}$ pointing into the bisector of the angle spread by $\vec{s}_1$ and $\vec{s}_2$:

$$\vec{s}_1 = \vec{s}_2 = \vec{s} \tag{2.79}$$

$\vec{b}_1$ and $\vec{b}_2$ are accordingly replaced by a unit vector $\vec{b}$ pointing into the bisector of the angle spread by $\vec{b}_1$ and $\vec{b}_2$

$$\vec{b}_1 = \vec{b}_2 = \vec{b} \tag{2.80}$$

The displacement vector $\vec{d}(x,y,z)$ is given by:

$$\vec{d} = \overrightarrow{P_1B} - \overrightarrow{P_2B} \tag{2.81}$$

and

$$\vec{d} = \overrightarrow{SP_2} - \overrightarrow{SP_1} \tag{2.82}$$

Inserting Eqs. (2.79) to (2.82) into Eq. (2.78) gives:

$$\delta = \left(\vec{b} - \vec{s}\right)\vec{d} \tag{2.83}$$

The following expression results for the interference phase:

$$\Delta\varphi(x,y) = \frac{2\pi}{\lambda}\vec{d}(x,y,z)\left(\vec{b} - \vec{s}\right) = \vec{d}(x,y,z)\vec{S} \tag{2.84}$$

The vector

$$\vec{S} = \frac{2\pi}{\lambda}\left(\vec{b} - \vec{s}\right) \tag{2.85}$$

is called the *sensitivity vector*. The sensitivity vector is only defined by the geometry of the holographic arrangement. It gives the direction in which the set-up has maximum sensitivity. At each point the projection of the displacement vector onto the sensitivity vector is measured. Equation (2.84) is the basis of all quantitative measurements of the deformation of opaque bodies.

In the general case of a three dimensional deformation field Eq. (2.84) contains the three components of $\vec{d}$ as unknown parameters. Three interferograms of the same surface with linear independent sensitivity vectors are necessary to determine the displacement. In many practical cases it is not the three dimensional displacement field that is of interest, but the deformation perpendicular to the surface. This *out-of-plane* deformation can be measured using an optimised set-up with parallel illumination and observation directions $(\vec{S} = 2\pi/\lambda(0,0,2))$. The component d_z is then calculated from the interference phase by

$$d_z = \Delta\varphi \frac{\lambda}{4\pi} \tag{2.86}$$

A phase variation of 2π corresponds to a deformation of $\lambda/2$.

2.7.3 Holographic Contouring

Another application of HI is the generation of a fringe pattern corresponding to contours of constant elevation with respect to a reference plane. Such contour fringes can be used to determine the shape of a three-dimensional object.

Holographic contour interferograms can be generated by different methods. In the following the

- *two-wavelength method* and the
- *two-illumination-point method*

are described. A third method, *the two-refractive-index technique*, has less practical applications and is not considered here.

The principal set-up of the two-wavelength method is shown in Fig. 2.19. A plane wave illuminates the object surface. The back scattered light interferes with the plane reference wave at the holographic recording medium. In the set-up of Fig. 2.19 the illumination wave is reflected onto the object surface via a beam splitter in order to ensure parallel illumination and observation directions. Two holograms are recorded with different wavelengths λ_1 and λ_2 on the same

Fig. 2.19 Holographic contouring

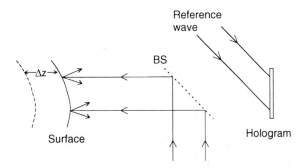

photographic plate. This can be done either simultaneously using two lasers with different wavelengths or in succession by changing the wavelength of a tuneable laser, e.g. a dye laser. After processing, the double exposed hologram is replaced and reconstructed with only one of the two wavelengths, say λ_2. Two virtual images become visible. The image recorded with λ_2 coincides with the object surface. The other image, recorded with λ_1 but reconstructed with λ_2, is slightly distorted. The z-coordinate of this image z' is calculated with the imaging Eq. (2.66):

$$z' = \frac{z_R^2 z}{z z_R + \frac{\lambda_2}{\lambda_1} z_R^2 - \frac{\lambda_2}{\lambda_1} z z_R} \approx z \frac{\lambda_1}{\lambda_2} \tag{2.87}$$

The indices "1" for virtual image $(z_1' \equiv z')$ and "O" for object $(z_O \equiv z)$ are omitted and it is assumed not to change the source coordinates of the reconstruction wave with respect to those of the recording coordinates $(z_P \equiv z_R \rightarrow \infty)$. The axial displacement of the image recorded with λ_1 but reconstructed with λ_2 is therefore:

$$\Delta z = z' - z = z \frac{|\lambda_1 - \lambda_2|}{\lambda_2} \tag{2.88}$$

The path difference of the light rays on their way from the source to the surface and from the surface to the hologram is $2\Delta z$. The corresponding phase shift is thus,

$$\Delta\varphi(x,y) = \frac{2\pi}{\lambda_1} 2\Delta z = 4\pi z \frac{|\lambda_1 - \lambda_2|}{\lambda_1 \lambda_2} \tag{2.89}$$

The two shifted images interfere. According to Eq. (2.89) the phase shift depends on the distance z from the hologram plane. All points of the object surface having the same z-coordinate (height) are therefore connected by a contour line. As a result an image of the surface superimposed by contour fringes develops. The height jump between adjacent fringes is:

$$\Delta H = z(\Delta\varphi = (n+1)2\pi) - z(\Delta\varphi = n2\pi) = \frac{\lambda_1\lambda_2}{2|\lambda_1 - \lambda_2|} = \frac{\Lambda}{2} \qquad (2.90)$$

$\Lambda = \lambda_1\lambda_2/|\lambda_1 - \lambda_2|$ is known as the *synthetic wavelength* or *equivalent wavelength*. The object is intersected by parallel planes which have a distance of ΔH, see the principle in Fig. 2.20 and a typical example in Fig. 2.21.

The equations derived in this chapter are valid only for small wavelength differences, because in addition to the axial displacement (which generates contour lines) also a lateral image displacement occurs. This lateral displacement can be neglected for small wavelength differences.

The principle of the two-illumination-point method is to make a double exposure hologram in which the point source illuminating the object is shifted slightly between the two exposures. If the illumination point S is shifted to S' between the two exposures (Fig. 2.22), the resulting optical path length difference δ is:

$$\delta = \overline{SP} + \overline{PB} - \left(\overline{S'P} + \overline{PB}\right) = \overline{SP} - \overline{S'P}$$
$$= \vec{s_1}\overrightarrow{SP} - \vec{s_2}\overrightarrow{S'P} \qquad (2.91)$$

The unit vectors $\vec{s_1}$ and $\vec{s_2}$ are defined as for the derivation of the interference phase due to deformation in Sect. 2.7.2. The same approximation is used and these vectors are replaced by a common unit vector:

$$\vec{s_1} = \vec{s_2} = \vec{s} \qquad (2.92)$$

Furthermore,

$$\vec{p} = \overrightarrow{SP} - \overrightarrow{S'P} \qquad (2.93)$$

is introduced as a vector from S to S'. The optical path difference is then given by

$$\delta = \vec{p}\,\vec{s} \qquad (2.94)$$

Fig. 2.20 Object intersection by contour lines

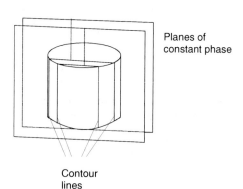

Planes of constant phase

Contour lines

Fig. 2.21 Two-wavelength contour fringes

Fig. 2.22 Two-illumination point contouring

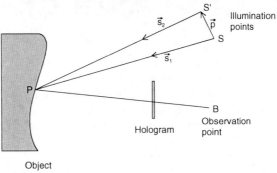

The corresponding phase change is:

$$\Delta\varphi = \frac{2\pi}{\lambda}\,\vec{p}\,\vec{s} \tag{2.95}$$

The object surface is intersected by fringes which consist of a set of hyperboloids. Their common foci are the two points of illumination S and S'. If the dimensions of the object are small compared to the distances between the source points and the object, plane contouring surfaces result. A collimated illumination together with a telecentric imaging system also generates plane contouring surfaces. The distance between two neighbouring surfaces is

$$\Delta H = \frac{\lambda}{2 \sin \frac{\theta}{2}} \tag{2.96}$$

where θ is the angle between the two illumination directions. Equation (2.96) is analogue to the fringe spacing in an interference pattern formed by two intersecting plane waves, see Eq. (2.29) in Sect. 2.2.

2.7.4 Refractive Index Measurement by HI

Another application of HI is the measurement of refractive index variations within transparent media. This mode of HI is used to determine temperature or concentration variations in fluid or gaseous media.

A refractive index change in a transparent medium causes a change of the optical path length and thereby a phase variation between two light waves passing the medium before and after the change. The interference phase due to refractive index variations is given by:

$$\Delta\varphi(x,y) = \frac{2\pi}{\lambda} \int_{l_1}^{l_2} [n(x,y,z) - n_0] dz \tag{2.97}$$

where n_0 is the refractive index of the medium under observation in its initial, unperturbed state and $n(x, y, z)$ is the final refractive index distribution. The light passes through the medium in the z-direction and integration is along the propagation direction. Equation (2.97) is valid for small refractive index gradients, where the light rays propagate along straight lines. The simplest case is that of a two-dimensional phase object with no variation of refractive index in z. In this case the refractive index distribution $n(x, y)$ can be calculated directly from Eq. (2.97). In the general case of a refractive index varying also in the z-direction Eq. (2.97) cannot be solved without further information about the process. However, in many practical experiments only two-dimensional phase objects have to be considered.

A set-up for the recording of holograms of transparent phase objects consists of a coherent light source, the transparent medium under investigation and optical components as in Fig. 2.23.

Fig. 2.23 Recording set-up
for transparent phase objects

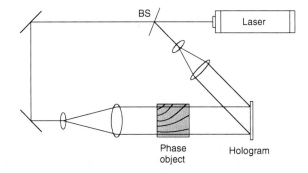

The laser beam is split into two separate waves. One wave is expanded by a telescopic lens system and illuminates the medium, which is located for example in a test cell with transparent walls. The transmitted part, the object wave, interferes with the reference wave at the surface of the hologram plate. After processing, the object wave is reconstructed by illuminating the hologram with the reference wave again, Fig. 2.24. Holographic Interferometry can be carried out either by the double exposure method or by the real-time method.

A holographic interferogram of a pure transparent object without any scattering consists of clear fringes undisturbed by speckle noise. These fringes are not localized in space, because there are no object contours visible. Yet, for some applications localized fringes are desired. In that case a diffusing screen can be placed in front of or behind the object volume.

2.7.5 Phase Shifting HI

As discussed in Sect. 2.7.1 it is not possible to calculate $\Delta\varphi$ unambiguously from the measured intensity, because the parameters $A(x, y)$ and $B(x, y)$ in Eq. (2.77) are not known and the sign is not determined.

Phase shifting Holographic Interferometry is a method which enables us to determine the interference phase by recording additional information [17, 36, 98,

Fig. 2.24 Reconstruction of
phase objects

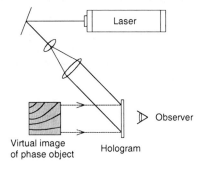

99]. The principle is to record three or more interference patterns with mutual phase shifts. For the case of three recordings, the interference patterns are described by:

$$I_1(x,y) = A(x,y) + B(x,y)\cos(\Delta\varphi)$$
$$I_2(x,y) = A(x,y) + B(x,y)\cos(\Delta\varphi + \alpha) \qquad (2.98)$$
$$I_3(x,y) = A(x,y) + B(x,y)\cos(\Delta\varphi + 2\alpha)$$

The equation system can be solved unambiguously for $\Delta\varphi$ if the phase angle α is known (e.g. 120°).

The phase shift can be realized in practice for example by employing a mirror mounted on a piezo-electric translator. The mirror is placed either in the object beam or in the reference beam. If appropriate voltages are applied to the piezo-electric translator during the hologram reconstruction, well defined path changes in the range of fractions of a wavelength can be introduced. These path changes correspond to phase differences between object—and reference wave.

Instead of using the minimum number of three reconstructions with two mutual phase shifts, Eq. (2.98), it is also possible to generate four reconstructions with three mutual phase shifts:

$$I_1(x,y) = A(x,y) + B(x,y)\cos(\Delta\varphi)$$
$$I_2(x,y) = A(x,y) + B(x,y)\cos(\Delta\varphi + \alpha)$$
$$I_3(x,y) = A(x,y) + B(x,y)\cos(\Delta\varphi + 2\alpha) \qquad (2.99)$$
$$I_4(x,y) = A(x,y) + B(x,y)\cos(\Delta\varphi + 3\alpha)$$

In that case the equation system can be solved without knowledge of the phase shift angle, α, as long as it is constant. The solution for $\Delta\varphi$ is [121]:

$$\Delta\varphi = \arctan\frac{\sqrt{I_1 + I_2 - I_3 - I_4}\cdot\sqrt{3I_2 - 3I_3 - I_1 + I_4}}{I_2 + I_3 - I_1 - I_4} \qquad (2.100)$$

Various HI phase shifting methods have been published [121], which differ in the number of recordings (at least 3), the value of α, and the method of generating the phase shift (stepwise or continuously). These methods will not be discussed in detail here. The principle has been described briefly in order to prepare for a comparison of phase determination in conventional HI using photographic plates and with the techniques used to obtain phase information in Digital Holographic Interferometry (Chap. 4). Finally it is noted that phase shifting HI is not the only way to determine the phase from a fringe pattern, but it is the most commonly applied. Other phase evaluating techniques include Fourier Transform methods, skeletonizing or heterodyne techniques.

2.7.6 *Phase Unwrapping*

Even after having determined the interference phase by a method such as phase shifting HI, a problem remains: the cosine function is periodic, i.e. the interference phase distribution is indefinite to an additive integer of 2π:

$$\cos(\Delta\varphi) = \cos(\Delta\varphi + 2\pi n)\, n \in Z \qquad (2.101)$$

Interference phase maps calculated with the arctan function or other inverse trigonometric functions therefore contain 2π jumps at those positions where an extreme value of $\Delta\varphi$ (either $-\pi$ or π) is reached. The interference phase change along a line of such a phase image resembles a saw tooth function, Fig. 2.25a. The correction of these modulo 2π jumps in order to generate a continuous phase distribution is called *demodulation, continuation* or *phase unwrapping*.

Several unwrapping algorithms have been developed in the last years. In the following the so called path-dependent unwrapping algorithm is described. At first a one-dimensional interference phase distribution is considered. The difference between the phase values of adjacent pixels $\Delta\varphi(n+1) - \Delta\varphi(n)$ is calculated. If this difference is less than $-\pi$, all phase values from the $(n+1)$th pixel onwards are increased by 2π. If this difference is greater than $+\pi$, 2π is subtracted from all phase values, starting from $(n+1)$. If none of the above mentioned conditions is valid the phase value remains unchanged. The practical implementation of this procedure is done by first calculating a step function, which cumulates the 2π jumps for all pixels, Fig. 2.25b. The continuous phase distribution is then calculated by adding

Fig. 2.25 Phase unwrapping.
a Interference phase modulo
2π: $\Delta\varphi_{2\pi}(x)$ **b** Step function:
$\Delta\varphi_{jump}(x)$ **c** unwrapped
interference phase:
$\Delta\varphi_{2\pi}(x) + \Delta\varphi_{jump}(x)$

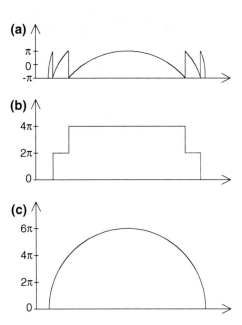

this step function to the unwrapped phase distribution, Fig. 2.25c. Almost every pixel can be used as a starting point for this unwrapping procedure, not necessarily the pixel at the start of the line. If a central pixel is chosen as the starting point the procedure has to be carried out in both directions from that point.

This one-dimensional unwrapping scheme can be transferred to two dimensions. One possibility is to unwrap first one row of the two dimensional phase map with the algorithm described above. The pixels of this unwrapped row act then as starting points for column demodulation.

One disadvantage of the simple unwrapping procedure described here is that difficulties occur if masked regions are in the phase image. These masked areas might be caused by e.g. holes in the object surface. To avoid this and other difficulties several other, more sophisticated demodulation algorithms have been developed [121].

Finally it should be mentioned that the unwrapping procedure is always the same for all methods of metrology that generate saw-tooth like images. This means the various unwrapping algorithm developed for HI and other methods can be used also for Digital Holographic Interferometry, because this technique also generates modulo 2π-images (see Chap. 4).

Chapter 3
Digital Holography

3.1 General Principles

The concept of digital holographic recording is illustrated in Fig. 3.1a [196, 198]. A plane reference wave and the wave reflected from the object interfere at the surface of an electronic sensor array (e.g. Charged Coupled Device, CCD, or Complementary Metal Oxide Semiconductor, CMOS). The resulting hologram is electronically recorded and stored in a computer. The object is, in general, a three dimensional body with diffusely reflecting surfaces, located at a distance d from the sensor (measured to some representative plane). This is just the classical off-axis geometry of photographic holography save that the recording medium is an electronic sensor array rather than photographic film.

In classical optical reconstruction using a replica of the original reference wave to illuminate the hologram, a "virtual" (primary) image is recreated at a distance d behind the sensor plane as viewed by an observer; a "real" (secondary) image is also formed at a distance d, from the sensor but in front of it, between it and the observer, see Fig. 3.1b. In DH, though, a physical image in virtual or real space is not created; numerical reconstruction by computer at a given plane produces a primary or secondary image on a monitor.

Using the coordinate system of Fig. 3.2, a light wave diffracted at an aperture (in this case a hologram) perpendicular to an incoming beam is described by the Fresnel-Kirchhoff integral, see Eq. (2.48), as

$$\Gamma(\xi',\eta') = \frac{i}{\lambda} \int\limits_{-\infty}^{\infty} \int\limits_{-\infty}^{\infty} h(x,y)E_R(x,y)\frac{\exp\left(-i\frac{2\pi}{\lambda}\rho'\right)}{\rho'}\,dxdy \qquad (3.1)$$

© Springer-Verlag Berlin Heidelberg 2015
U. Schnars et al., *Digital Holography and Wavefront Sensing*,
DOI 10.1007/978-3-662-44693-5_3

Fig. 3.1 Digital Holography.
a Recording,
b Reconstruction with
reference wave E_R,
c Reconstruction with
conjugate reference wave E_R^*

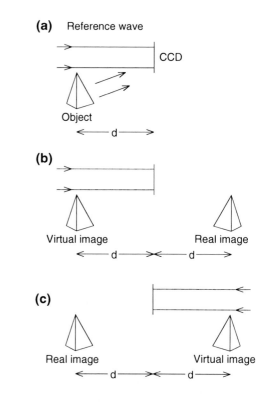

Fig. 3.2 Coordinate system
for numerical hologram
reconstruction

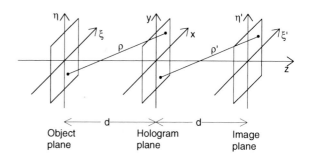

where

$$\rho' = \sqrt{(x - \xi')^2 + (y - \eta')^2 + d^2} \qquad (3.2)$$

$h(x,y)$ is the hologram function and ρ' is the distance between a point in the
hologram plane and a point in the reconstruction plane. The inclination factor is set
to 1, since the angles θ' and θ'' (see Sect. 2.4) are approximately zero. This is valid
for all the numerical reconstruction algorithms in this book.

A plane reference wave $E_R(x, y)$ can be described in terms of its real amplitude,

$$E_R = a_R + i0 = a_R \tag{3.3}$$

The diffraction pattern is calculated at a distance d behind the sensor plane, so that it reconstructs the complex amplitude of the wave in the plane of the real image.

Equation (3.1) forms the basis for numerical reconstruction from a hologram. Because the reconstructed wave field $\Gamma(\xi', \eta')$ is a complex function, both the intensity as well as its phase can be extracted [195]. This is in contrast to the case of optical hologram reconstruction, in which only the intensity is obtainable. This interesting property of Digital Holography is used in Digital Holographic Interferometry, see Chap. 4.

As discussed in Sect. 2.6 the real image could be distorted. According to Eq. (2.63) an undistorted real image can be produced by using the conjugate reference beam for reconstruction. To numerically reconstruct an undistorted real image it is therefore necessary to insert E_R^* instead of E_R into Eq. (3.1):

$$\Gamma(\xi, \eta) = \frac{i}{\lambda} \int\limits_{-\infty}^{\infty} \int\limits_{-\infty}^{\infty} h(x, y) E_R^*(x, y) \frac{\exp\left(-i\frac{2\pi}{\lambda}\rho\right)}{\rho} dx dy \tag{3.4}$$

with

$$\rho = \sqrt{(x - \xi)^2 + (y - \eta)^2 + d^2} \tag{3.5}$$

This reconstruction scheme is shown in Fig. 3.1c. The real image is formed at the position where the object was located during recording. It should be noted that for the plane reference wave defined in Eq. (3.3) both reconstruction formulas, Eqs. (3.1) and (3.4), are equivalent since $E_R = E_R^* \equiv a_R$.

The arrangement of Fig. 3.1 with a plane reference wave perpendicularly illuminating the sensor is commonly used in Digital Holography. Other recording geometries are discussed later.

Reconstruction of the virtual image is also possible by either selecting the negative branch of the square root or introducing the imaging properties of a lens into the numerical reconstruction process [196]. This lens corresponds to the eye of an observer viewing an optically reconstructed hologram. In the simplest case this lens is located directly behind the hologram, as in Fig. 3.3. The imaging properties of a lens with focal distance f are represented by a complex factor, $L(x, y)$, as

$$L(x, y) = \exp\left[i\frac{\pi}{\lambda f} (x^2 + y^2)\right] \tag{3.6}$$

Fig. 3.3 Reconstruction of
the virtual image

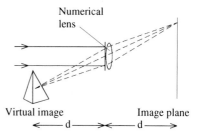

The factor $L(x,y)$ is calculated in Annex B1. For unity magnification, the lens should have a focal length of $f = d/2$.

The lens described by Eq. (3.6) introduces phase aberrations, which can be corrected by multiplying the reconstructed wave field by another factor

$$P(\xi',\eta') = \exp\left[i\frac{\pi}{\lambda f}\left(\xi'^2 + \eta'^2\right)\right] \tag{3.7}$$

This correction factor is derived in Annex B2. The full equation which describes reconstruction via a virtual lens with $f = d/2$ is therefore

$$\Gamma(\xi',\eta') = \frac{i}{\lambda}P(\xi',\eta')\int\limits_{-\infty}^{\infty}\int\limits_{-\infty}^{\infty} h(x,y)E_R(x,y)L(x,y)\frac{\exp\left(-i\frac{2\pi}{\lambda}\rho'\right)}{\rho'}dxdy \tag{3.8}$$

3.2 Numerical Reconstruction

3.2.1 Reconstruction by the Fresnel Approximation

For x- and y-values, as well as for ξ- and η-values, which are small compared to the distance d between the reconstruction plane and the sensor, the expression Eq. (3.5) can be expanded with a Taylor series:

$$\rho = d + \frac{(\xi - x)^2}{2d} + \frac{(\eta - y)^2}{2d} - \frac{1}{8}\frac{\left[(\xi - x)^2 + (\eta - y)^2\right]^2}{d^3} + \cdots \tag{3.9}$$

The fourth term in Eq. (3.9) can be neglected, if it is small compared to the wavelength [116], i.e. if,

$$\frac{1}{8}\frac{\left[(\xi - x)^2 + (\eta - y)^2\right]^2}{d^3} \ll \lambda \tag{3.10}$$

Or, rewriting in terms of d, we have

$$d \gg \sqrt[3]{\frac{1}{8} \frac{[(\xi - x)^2 + (\eta - y)^2]^2}{\lambda}} \tag{3.11}$$

Then the distance ρ consists of linear and quadratic terms:

$$\rho = d + \frac{(\xi - x)^2}{2d} + \frac{(\eta - y)^2}{2d} \tag{3.12}$$

With the additional approximation of replacing the denominator in (3.4) by d the following expression results for reconstruction of the real image:

$$\Gamma(\xi, \eta) = \frac{i}{\lambda d} \exp\left(-i\frac{2\pi}{\lambda} d\right)$$
$$\times \int_{-\infty}^{\infty} \int_{-\infty}^{\infty} E_R^*(x, y) h(x, y) \exp\left[-i\frac{\pi}{\lambda d}\left((\xi - x)^2 + (\eta - y)^2\right)\right] dxdy \tag{3.13}$$

If the multiplication terms in the argument of the exponential under the integral are carried out we get

$$\Gamma(\xi, \eta) = \frac{i}{\lambda d} \exp\left(-i\frac{2\pi}{\lambda} d\right) \exp\left[-i\frac{\pi}{\lambda d}(\xi^2 + \eta^2)\right]$$
$$\times \int_{-\infty}^{\infty} \int_{-\infty}^{\infty} E_R^*(x, y) h(x, y) \exp\left[-i\frac{\pi}{\lambda d}(x^2 + y^2)\right] \exp\left[i\frac{2\pi}{\lambda d}(x\xi + y\eta)\right] dxdy \tag{3.14}$$

This equation is known as the *Fresnel approximation* or *Fresnel transformation* due to its mathematical similarity with the Fourier Transform (see below). It enables reconstruction of the wavefield in a plane behind the hologram, in this case in the plane of the real image.

The intensity is given by its square,

$$I(\xi, \eta) = |\Gamma(\xi, \eta)|^2 \tag{3.15}$$

and its phase by

$$\varphi(\xi, \eta) = \arctan\frac{\text{Im}[\Gamma(\xi, \eta)]}{\text{Re}[\Gamma(\xi, \eta)]} \tag{3.16}$$

where "Re" denotes the real part and "Im" the imaginary part of the wave.

Reconstruction of the virtual image in the Fresnel approximation can be expressed as,

$$
\Gamma(\xi', \eta') = \frac{i}{\lambda d} \exp\left(-i\frac{2\pi}{\lambda}d\right) \exp\left[-i\frac{\pi}{\lambda d}(\xi'^2 + \eta'^2)\right] P(\xi', \eta')
$$

$$
\times \int_{-\infty}^{\infty} \int_{-\infty}^{\infty} E_R(x,y)L(x,y)h(x,y) \exp\left[-i\frac{\pi}{\lambda d}(x^2 + y^2)\right] \exp\left[i\frac{2\pi}{\lambda d}(x\xi' + y\eta')\right] dxdy
$$

$$
= \frac{i}{\lambda d} \exp\left(-i\frac{2\pi}{\lambda}d\right) \exp\left[+i\frac{\pi}{\lambda d}(\xi'^2 + \eta'^2)\right]
$$

$$
\times \int_{-\infty}^{\infty} \int_{-\infty}^{\infty} E_R(x,y)h(x,y) \exp\left[+i\frac{\pi}{\lambda d}(x^2 + y^2)\right] \exp\left[i\frac{2\pi}{\lambda d}(x\xi' + y\eta')\right] dxdy
$$

$$(3.17)$$

Alternatively we can insert a negative distance into Eq. (3.14), which has the added advantage that the virtual image is not rotated by 180° because of the action of performing the Fourier transform.

To digitise the Fresnel transform in Eq. (3.14), the following definitions and substitutions are introduced [261],

$$
u = \frac{\xi}{\lambda d}; \quad v = \frac{\eta}{\lambda d} \tag{3.18}
$$

Thus (3.14) is now expressed as,

$$
\Gamma(u,v) = \frac{i}{\lambda d} \exp\left(-i\frac{2\pi}{\lambda}d\right) \exp\left[-i\pi\lambda d(u^2 + v^2)\right]
$$

$$
\times \int_{-\infty}^{\infty} \int_{-\infty}^{\infty} E_R^*(x,y)h(x,y) \exp\left[-i\frac{\pi}{\lambda d}(x^2 + y^2)\right] \exp[i2\pi(xu + yv)]dxdy
$$

$$(3.19)$$

A comparison of Eq. (3.19) with the definition of the two-dimensional Fourier transform (see Annex A) shows that the Fresnel approximation is the just the inverse Fourier transformation of the function $E_R^*(x,y)h(x,y) \exp[-i\pi/\lambda d(x^2 + y^2)]$,

$$
\Gamma(u,v) = \frac{i}{\lambda d} \exp\left(-i\frac{2\pi}{\lambda}d\right) \exp\left[-i\pi\lambda d(u^2 + v^2)\right]
$$

$$
\times \Im^{-1}\left\{ E_R^*(x,y)h(x,y) \exp\left[-i\frac{\pi}{\lambda d}(x^2 + y^2)\right]\right\} \tag{3.20}
$$

The function Γ can be digitised if the hologram function $h(x,y)$ is sampled on a rectangular raster of N × N points, with steps Δx and Δy along the coordinates. The distances between neighbouring pixels on the sensor array in the horizontal and vertical directions are given by Δx and Δy respectively. With these discrete values included, the integrals in (3.19) are written in terms of finite sums, i.e.

$$\Gamma(m,n) = \frac{i}{\lambda d} \exp\left(-i\frac{2\pi}{\lambda}d\right) \exp\left[-i\pi\lambda d\left(m^2\Delta u^2 + n^2\Delta v^2\right)\right]$$

$$\times \sum_{k=0}^{N-1}\sum_{l=0}^{N-1} E_R^*(k,l)h(k,l)\exp\left[-i\frac{\pi}{\lambda d}\left(k^2\Delta x^2 + l^2\Delta y^2\right)\right]\exp[i2\pi(k\Delta xm\Delta u + l\Delta yn\Delta v)]$$

$$\text{for } m = 0,1,...,N-1; \text{ and } n = 0,1,...,N-1$$

(3.21)

According to Fourier transform procedures, and Δu, Δv can be written in terms of Δx, Δy (see Annex A) as,

$$\Delta u = \frac{1}{N\Delta x}; \quad \Delta v = \frac{1}{N\Delta y} \tag{3.22}$$

After re-substitution, we have,

$$\Delta\xi = \frac{\lambda d}{N\Delta x}; \quad \Delta\eta = \frac{\lambda d}{N\Delta y} \tag{3.23}$$

Applying these relationships, Eq. (3.21) converts to

$$\Gamma(m,n) = \frac{i}{\lambda d}\exp\left(-i\frac{2\pi}{\lambda}d\right)\exp\left[-i\pi\lambda d\left(\frac{m^2}{N^2\Delta x^2} + \frac{n^2}{N^2\Delta y^2}\right)\right]$$

$$\times \sum_{k=0}^{N-1}\sum_{l=0}^{N-1} E_R^*(k,l)h(k,l)\exp\left[-i\frac{\pi}{\lambda d}\left(k^2\Delta x^2 + l^2\Delta y^2\right)\right]\exp\left[i2\pi\left(\frac{km}{N} + \frac{ln}{N}\right)\right]$$

(3.24)

This is the discrete Fresnel transform. The matrix Γ is evaluated by multiplying $E_R^*(k,l)$ with $h(k,l)$ and $\exp[-i\pi/(\lambda d)(k^2\Delta x^2 + l^2\Delta y^2)]$, followed by application of an inverse discrete Fourier transform to the product. This calculation is accomplished most efficiently using the Fast Fourier Transform (FFT) algorithm. The factors before the sum term in Eq. (3.24) only affect the overall phase and can be neglected if it is only the intensity in Eq. (3.15) that is of interest. This is also the case if phase differences between holograms recorded with the same wavelength have to be calculated, according to,$(\Delta\varphi = \varphi_1 + const. - (\varphi_2 + const.) = \varphi_1 - \varphi_2)$.

The corresponding discrete formula for reconstruction with a virtual lens of $f = d/2$ (Eq. 3.17) is,

$$\Gamma(m,n) = \frac{i}{\lambda d}\exp\left(-i\frac{2\pi}{\lambda}d\right)\exp\left[+i\pi\lambda d\left(\frac{m^2}{N^2\Delta x^2}+\frac{n^2}{N^2\Delta y^2}\right)\right]$$
$$\times \sum_{k=0}^{N-1}\sum_{l=0}^{N-1}E_R(k,l)h(k,l)\exp\left[+i\frac{\pi}{\lambda d}\left(k^2\Delta x^2+l^2\Delta y^2\right)\right]\exp\left[i2\pi\left(\frac{km}{N}+\frac{ln}{N}\right)\right]$$

(3.25)

A typical digital hologram of a dice recorded with the geometry of Fig. 3.1 is shown in Fig. 3.4. The dice is placed a distance $d = 1.054$ m from a sensor array with $1,024 \times 1,024$ pixels of pitch $\Delta x = \Delta y = 6.8$ μm. The recording wavelength is 632.8 nm. Numerical reconstruction of the real image is performed according to Eqs. (3.14) and (3.24) and illustrated in Fig. 3.5. The bright square in the centre of the image is the non-diffracted (zero order) reconstruction wave and corresponds to the first term on the right side of Eq. (2.63). Because of the off-axis geometry, the image is spatially separated from the zero order term. The other (virtual) image is out-of-focus in this reconstruction.

An interesting property of (off-axis) holography is that every part of a hologram contains all the information about the entire object. This is illustrated by the holograms of Figs. 3.6 and 3.8, where black masks cover nearly half of the hologram areas. Nevertheless, the entire cube is visible without obstruction in the reconstructions (Figs. 3.7 and 3.9). The masks are visible as shadows in the zero order terms. The reduction of the effective pixel number leads to a consequent

Fig. 3.4 Digital hologram of a die

Fig. 3.5 Numerical
reconstruction

Fig. 3.6 Masked digital
hologram

reduction of the resolution in the reconstructed images. This is equivalent to the
increase of the speckle size observed in optical hologram reconstruction when the
aperture is reduced.

Regarding Eq. (3.23), the pixel distances in the reconstructed image $\Delta\xi$ and $\Delta\eta$
are dependent on the chosen numerical reconstruction distance d. This is because

Fig. 3.7 Reconstruction

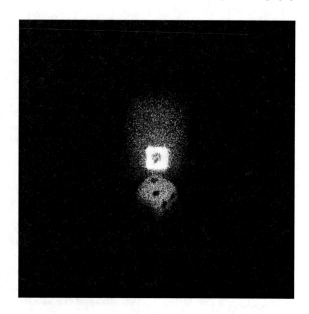

Fig. 3.8 Masked digital
hologram

Eq. (3.23) corresponds to the diffraction limited resolution of an optical system. The
hologram corresponds to the aperture of the optical system with a side of length
$N\Delta x$; a diffraction pattern develops at a distance d behind the hologram. The term
$\Delta \xi = \lambda d / N\Delta x$ therefore describes the half-diameter of the Airy disk or the speckle
diameter in the plane of the reconstructed image, accordingly, limits the resolution.

Fig. 3.9 Reconstruction

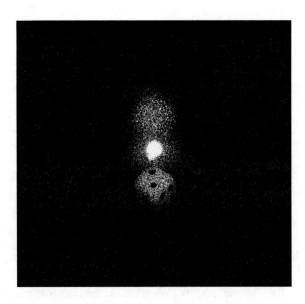

This can be regarded as a "natural scaling" algorithm, setting the resolution of the image reconstructed by a discrete Fresnel transform always to the physical limit.

A simple Matlab© Fresnel transformation reconstruction routine is shown in Appendix C.

3.2.2 Reconstruction by the Convolution Approach

Numerical processing of the Fresnel-Kirchhoff integral Eqs. (3.1) and (3.4) without the application of any approximations is time consuming. For faster and more efficient numerical processing, a different but equivalent formulation is often more suitable. This formulation makes use of the convolution theorem and, within the scope of this book, is accordingly denoted as the "convolution approach". Some other publications use the term *Angular Spectrum Method* (ASM), see e.g. [243]. Demetrakopoulos and Mittra applied this method for numerical reconstruction of suboptical holograms [41]. Later this approach was applied to optical holography by Kreis [124].

The reconstruction formula Eq. (3.4) can be interpreted as a superposition integral,

$$\Gamma(\xi, \eta) = \int\limits_{-\infty}^{\infty} \int\limits_{-\infty}^{\infty} h(x, y) E_R^*(x, y) g(\xi, \eta, x, y) dx dy \qquad (3.26)$$

where the impulse response $g(x,y,\xi,\eta)$ is given by

$$g(\xi,\eta,x,y) = \frac{i}{\lambda} \frac{\exp\left[-i\frac{2\pi}{\lambda}\sqrt{d^2 + (x-\xi)^2 + (y-\eta)^2}\right]}{\sqrt{d^2 + (x-\xi)^2 + (y-\eta)^2}} \tag{3.27}$$

According to Eq. (3.26) the linear system characterized by $g(\xi,\eta,x,y) = g(\xi-x,\eta-y)$ is shift-invariant. The superposition integral can be regarded therefore as a convolution and the convolution theorem (Annex A) can be applied. According to this approach the Fourier transform of the convolution of $h \cdot E_R^*$ with g is the product of the individual transforms $\Im\{hE_R^*\}$ and $\Im\{g\}$. So $\Gamma(\xi,\eta)$ can be calculated by, firstly Fourier transforming $h \cdot E_R^*$, followed by multiplication with the Fourier transform of g, and, finally, taking an inverse Fourier transform of the product. Three Fourier transforms are therefore necessary to complete the whole process. The individual Fourier transforms are efficiently carried out using the FFT algorithm.

For numerical processing the discrete impulse response function has to be calculated, by replacing the continuous differences $(x-\xi)$ and $(y-\eta)$ with the discrete variables $k\Delta x$ and $l\Delta y$, thus

$$g(k,l) = \frac{i}{\lambda} \frac{\exp\left[-i\frac{2\pi}{\lambda}\sqrt{d^2 + k^2\Delta x^2 + l^2\Delta y^2}\right]}{\sqrt{d^2 + k^2\Delta x^2 + l^2\Delta y^2}} \tag{3.28}$$

with integer values $k = 0, 1, \ldots, N-1;\ l = 0, 1, \ldots, N-1$

The process of reconstruction into the real image plane can be written as,

$$\Gamma(\xi,\eta) = \Im^{-1}\left\{\Im\left(h \cdot E_R^*\right) \cdot \Im(g)\right\} \tag{3.29}$$

A simple Matlab© reconstruction routine perform Eq. (3.29) is shown in Appendix C.

The Fourier transform of $g(\xi,\eta,x,y)$ can be calculated and expressed analytically [130] as,

$$G(f_x,f_y) = \exp\left(-i\frac{2\pi d}{\lambda}\sqrt{1 - \lambda^2 f_x^2 - \lambda^2 f_y^2}\right) \tag{3.30}$$

The spatial frequencies f_x and f_y can now be replaced by discrete values,

$$f_x = \frac{n}{N\Delta x} \quad f_y = \frac{m}{N\Delta y} \tag{3.31}$$

with integer values $n = 0, 1, \ldots, N-1;\ m = 0, 1, \ldots, N-1$. The discrete transfer function G now becomes

$$G(n, m) = \exp\left\{ -i\frac{2\pi d}{\lambda}\sqrt{1 - \left(\frac{\lambda n}{N\Delta x}\right)^2 - \left(\frac{\lambda m}{N\Delta y}\right)^2} \right\} \tag{3.32}$$

which consequently saves one Fourier transform operation in reconstruction. Thus we now have,

$$\Gamma(\xi, \eta) = \Im^{-1}\left\{\Im\left(h \cdot E_R^*\right) \cdot G\right\} \tag{3.33}$$

To reconstruct the virtual image either a negative distance, or a lens with transmission factor $L(x, y)$ and a correction factor $P(\xi', \eta')$ according to Eqs. (3.6) and (3.7) have to be taken into account. Thus, we have

$$\Gamma(\xi', \eta') = P(\xi', \eta')\Im^{-1}\left\{\Im\left(h \cdot E_R \cdot L\right) \cdot G\right\} \tag{3.34}$$

The pixel spacing corresponding to the images reconstructed by the convolution approach are equal to that of the hologram pitch, i.e.

$$\Delta\xi = \Delta x; \quad \Delta\eta = \Delta y \tag{3.35}$$

The pixel separations in the reconstructed images corresponding to the convolution approach differ from those which occur with the Fresnel approximation (Eq. 3.23). At first sight it seems to be possible to achieve a higher resolution with the convolution approach if the pixel separation is small enough. However, on closer examination we recognise that the resolution calculated by Eq. (3.35) is only a numerical value. The physical image resolution is determined by the diffraction limit, i.e. Eq. (3.23) and this also applies to the resolution limit corresponding to the convolution approach.

The area reconstructed with the impulse response function defined in Eq. (3.32) is symmetrical with respect to the optical axis. The area can be shifted by introducing the integers s_k, l_l,

$$g(k + s_k, l + s_l) = \frac{i}{\lambda}\frac{\exp\left[-i\frac{2\pi}{\lambda}\sqrt{d^2 + (k + s_k)^2\Delta x^2 + (l + s_l)^2\Delta y^2}\right]}{\sqrt{d^2 + (k + s_k)^2\Delta x^2 + (l + s_l)^2\Delta y^2}} \tag{3.36}$$

The convolution approach allows us to introduce image magnification into the reconstruction process. This is possible if the reconstruction distance is set to

$$d' = d \cdot m \tag{3.37}$$

where d is the recording distance (also used as the reconstruction distance) and m is the magnification factor. A magnification of $m = 1$ corresponds to $\Delta\xi = \Delta x$, and $\Delta\eta = \Delta y$. The lens focal distance is given by the lens formula of geometrical optics:

Fig. 3.10 Reconstruction
with the convolution
approach

$$f = \left(\frac{1}{d} + \frac{1}{d'}\right)^{-1} \tag{3.38}$$

Now Eq. (3.34) is applied for reconstruction at a distance d' instead of d and,
thus

$$L(x, y) = \exp\left[i\frac{\pi}{\lambda f}(x^2 + y^2)\right] = \exp\left[i\frac{\pi}{\lambda}\left(\frac{1}{d} + \frac{1}{d'}\right)(x^2 + y^2)\right] \tag{3.39}$$

An example of a reconstruction with the convolution approach is shown in
Fig. 3.10. The hologram of Fig. 3.4 is reconstructed with a magnification of
$m = 1/7$. The corresponding pixel separation in the reconstructed image for Δx of
6.8 μm is given as $\Delta \xi = \Delta x/m = 48$ μm. This should be compared with $\Delta \xi =$
96 μm obtained using the Fresnel reconstruction (and shown in Fig. 3.5). Thus
twice as many pixels are available for the object field using the convolution
approach. However, it is emphasized again that the physical resolution is the same
in both Figs. 3.5 and 3.10.

3.2.3 Digital Fourier Holography

The special holographic recording geometry of Fig. 3.11 is known as *lensless
Fourier holography*. It also has been realized using digital holographic concepts

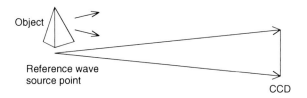

Fig. 3.11 Digital lensless Fourier holography

[245]. Here, a point source spherical reference wave is located in the plane of the object. The reference wave at the sensor plane is therefore described by,

$$E_R = \frac{\exp\left(-i\frac{2\pi}{\lambda}\sqrt{(d^2 + x^2 + y^2)}\right)}{\sqrt{(d^2 + x^2 + y^2)}}$$

$$\approx \frac{1}{d}\exp\left(-i\frac{2\pi}{\lambda}d\right)\exp\left(-i\frac{\pi}{\lambda d}(x^2 + y^2)\right) \tag{3.40}$$

The term $\sqrt{d^2 + x^2 + y^2}$ is the distance between the point source and a point with coordinates (x,y) in the sensor plane. The approximation in Eq. (3.40) is the same as used in Sect. 3.2.1 to derive the Fresnel transform. Inserting this expression into the reconstruction formula for the virtual image (3.17) leads to following equation,

$$\Gamma(\xi,\eta) = C\exp\left[+i\frac{\pi}{\lambda d}(\xi^2 + \eta^2)\right]\mathfrak{I}^{-1}\{h(x,y)\} \tag{3.41}$$

where C is a complex constant. A lensless Fourier hologram is therefore reconstructed by a Fourier transform. The spherical phase factor $\exp(-i\pi/\lambda d(x^2 + y^2))$ associated with the Fresnel transform is eliminated by the use of a spherical reference wave with the same curvature as the original. Numerical focusing into other planes is therefore not possible using Eq. (3.41). Numerical focusing can be realized, if different values of d for *recording* (reference wave factor E_R) and *reconstruction* are inserted in Eq. (3.17).

3.3 Shift and Suppression of DC-Term and Conjugate Image

3.3.1 Suppression of the DC Term

The bright square in the centre of Fig. 3.5 is the non-diffracted reconstruction wave. This zero order or DC term disturbs the image, because it obscures all the parts of the object which lie behind it. Methods have been developed to suppress this term e. g. by Kreis et al. [125].

To understand the origins of this DC term, the process of hologram formation as described by Eq. (2.60) needs to be considered again. The equation is rewritten by inserting the definitions of E_R and E_O, and multiplying, to give,

$$\begin{aligned} I(x,y) &= |E_0(x,y) + E_R(x,y)|^2 \\ &= a_R^2 + a_O^2 + 2a_R a_O \cos(\varphi_O - \varphi_R) \end{aligned} \tag{3.42}$$

The first two terms lead to the DC term in the reconstruction process. The third term is a cosinusoidally varying component lying between values of $\pm 2a_R a_O$ and illuminating the pixels across the sensor. The average intensity of all pixels of the hologram matrix is

$$I_m = \frac{1}{N^2} \sum_{k=0}^{N-1} \sum_{l=0}^{N-1} I(k\Delta x, l\Delta y) \tag{3.43}$$

The term $a_R^2 + a_O^2$ can now be suppressed by subtracting this average intensity I_m from the hologram:

$$I'(k\Delta x, l\Delta y) = I(k\Delta x, l\Delta y) - I_m(k\Delta x, l\Delta y) \tag{3.44}$$

for $k = 0, \ldots, N-1;\quad l = 0, \ldots, N-1$.

The reconstruction of I' creates an image with strongly suppressed DC term. An example of this is shown in Fig. 3.12. The upper left figure is a photograph of the object. Reconstruction without DC term suppression is depicted in the upper right figure. The object is covered by the DC term. The lower left figure shows reconstruction with DC suppression included. The original object is clearly visible.

Instead of subtracting the average intensity it is also possible remove the DC component using a high-pass filter with a low cut-off frequency as shown in the lower right image of Fig. 3.12.

The subtraction of the average intensity from the hologram before reconstruction is the basic objective of DC suppression. The same effect can be achieved, if two holograms with stochastically changed speckle structures are subtracted from each other [42]. The reconstruction of this subtraction hologram results in an image without zero order term.

Another method of suppression is to separately capture and measure the intensities of the reference wave a_R^2 and object wave a_O^2. This can be done for example by blocking one wave while monitoring the other. Afterwards a DC term free image can be calculated by subtracting the intensities from the hologram before reconstruction. However, this requires higher experimental effort due to the additional measurements needed.

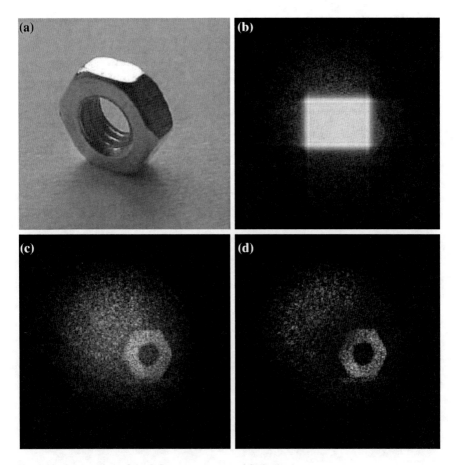

Fig. 3.12 Suppression of the DC term (courtesy of S. Seebacher)

3.3.2 Tilted Reference Wave

Using the recording geometry of Fig. 3.1 the real and virtual image are located at different observation planes. During numerical reconstruction we can choose to focus either on the real or on the virtual image. The other image is usually out-of-focus due to the long distance between the object and sensor. Consequently only one image is clearly visible in the reconstruction, see Fig. 3.5.

However, there are some instances where it is beneficial to laterally shift one image with respect to the other. In this case it can be useful to record the holograms with a tilted reference wave, as in Fig. 3.13. In this geometry the real image is deflected from the optic axis at an angle approximately twice that of the original reference wave.

Fig. 3.13 Digital Holography
with a tilted reference wave.
a Recording.
b Reconstruction

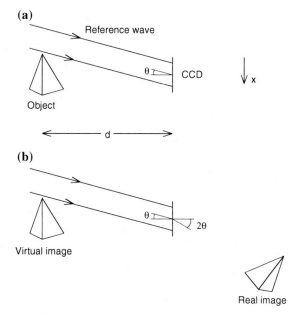

The tilted reference wave is described by,

$$E_R = \exp\left(-i\frac{2\pi}{\lambda}x\sin\theta\right) \tag{3.45}$$

The disadvantage of this set-up are the much higher spatial frequencies produced at the sensor in comparison to the geometry of Fig. 3.1.

3.3.3 Phase Shifting Digital Holography

The amplitude and phase of a light wave can be reconstructed from a single hologram by the methods described in the preceding chapters. A completely different approach, called Phase Shifting Digital Holography, has been proposed by Skarman [216, 251]. He used a phase shifting algorithm to calculate the *initial* phase and thus the complex amplitude in any plane, e.g. the image plane. With the initial complex amplitude distribution in one plane the wave field in any other plane can be determined using the Fresnel-Kirchhoff formulation. Later Phase Shifting DH was improved and applied to opaque objects by Yamaguchi et al. [92, 255–258, 264, 265].

The basic arrangement for phase shifting DH is shown in Fig. 3.14. The object wave and the reference wave interfere at the surface of a sensor. The reference wave is guided via a mirror mounted on a piezoelectric transducer (PZT). With this PZT

Fig. 3.14 Phase shifting
Digital Holography, set-up

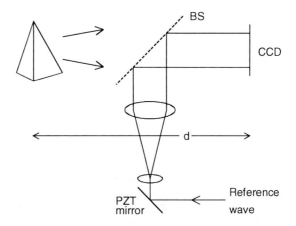

the phase of the reference wave can be shifted stepwise. Several (at least three) interferograms with mutual phase shifts are recorded. Afterwards the object phase φ_O is calculated from the phase shifted interferograms; the procedure is similar to that of phase shifting in conventional HI (see Sect. 2.7.5). The real amplitude $a_O(x, y)$ of the object wave can be extracted from the intensity by blocking the reference wave.

As a result the complex amplitude

$$E_O(x, y) = a_O(x, y) \exp(+i\varphi_O(x, y)) \tag{3.46}$$

of the object wave is determined in the recording (x, y) plane.

Now the Fresnel-Kirchhoff integral can be used to calculate the complex amplitude in any other plane. To calculate an image of the object an artificial lens with, for example, $f = d/2$ is introduced in the recording plane according to Eq. (3.6). By means of the Fresnel approximation Eq. (3.17) the complex amplitude in the image plane is then given by

$$
\begin{aligned}
E_O(\xi', \eta') &= C \exp\left[+i\frac{\pi}{\lambda d}\left(\xi'^2 + \eta'^2\right)\right] \\
&\quad \times \int\limits_{-\infty}^{\infty} \int\limits_{-\infty}^{\infty} E_O(x, y)L(x, y) \exp\left[-i\frac{\pi}{\lambda d}\left(x^2 + y^2\right)\right] \exp\left[i\frac{2\pi}{\lambda d}\left(x\xi' + y\eta'\right)\right] dxdy \\
&= C \exp\left[+\frac{i\pi}{\lambda d}\left(\xi'^2 + \eta'^2\right)\right] \\
&\quad \times \int\limits_{-\infty}^{\infty} \int\limits_{-\infty}^{\infty} E_O(x, y) \exp\left[+i\frac{\pi}{\lambda d}\left(x^2 + y^2\right)\right] \exp\left[i\frac{2\pi}{\lambda d}\left(x\xi' + y\eta'\right)\right] dxdy
\end{aligned}
\tag{3.47}
$$

where again the coordinate system of Fig. 3.2 applies. Since the complex amplitude in the hologram plane is known, it is also possible to reconstruct the object by inversion of the propagation process [206]. Propagation from the object plane to the hologram plane is described by

$$E_O(x,y) = \frac{i}{\lambda} \int_{-\infty}^{\infty} \int_{-\infty}^{\infty} E_O(\xi,\eta) \frac{\exp\left(-i\frac{2\pi}{\lambda}\sqrt{d^2 + (\xi - x)^2 + (\eta - y)^2}\right)}{\sqrt{d^2 + (\xi - x)^2 + (\eta - y)^2}} d\xi d\eta \qquad (3.48)$$

$$= \Im^{-1}\{\Im(E_O(\xi,\eta)) \cdot \Im(g(\xi,\eta,x,y))\}$$

with

$$g(\xi,\eta,x,y) = \frac{i}{\lambda} \frac{\exp\left(-i\frac{2\pi}{\lambda}\sqrt{d^2 + (\xi - x)^2 + (\eta - y)^2}\right)}{\sqrt{d^2 + (\xi - x)^2 + (\eta - y)^2}} \qquad (3.49)$$

$E_O(\xi,\eta)$ describes the complex amplitude of the object wave at the surface, see Fig. 3.2. Therefore it can be calculated directly by inverting Eq. (3.46), to give,

$$E_O(\xi,\eta) = \Im^{-1}\left\{\frac{\Im(E_O(x,y))}{\Im(g(\xi,\eta,x,y))}\right\} \qquad (3.50)$$

The advantage of phase shifting Digital Holography is that it produces a reconstructed image of the object without the presence of either the zero order term or the conjugate image. The price for this achievement is the higher technical effort required; phase shifted interferograms have to be generated, thereby restricting the method to slowly varying phenomena with constant phase during the recording cycle.

Phase shifting Digital Holography is illustrated by a holographic image of a nut, shown in Fig. 3.15. This example demonstrates the improvement compared to conventional Digital Holography, as shown in Fig. 3.12.

3.4 Recording of Digital Holograms

3.4.1 Image Sensors

It was the invention of the Charge-Coupled Device (CCD) at Bell Labs, and the dramatic increase in computer storage and processing power which led to the advent of digital holography. More recently, the Complementary Metal Oxide Semiconductor (CMOS) has also become popular for image sensing applications

Fig. 3.15 Phase shifting
Digital Holography, example
(courtesy of S. Seebacher)

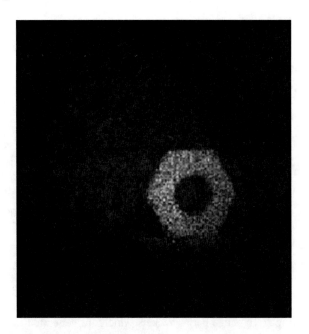

and is gradually replacing the CCD in digital still and video cameras. Electronic sensors like CCD or CMOS are composed of arrays of individual light sensitive elements (pixels) which convert incident photons into an induced charge, proportional to the incident intensity that can be stored or transferred through the device. The arrays are sensitive to the spatial variance of the incident light, and are therefore widely used in image recording. Both CCD and CMOS arrays are used in DH. CCDs are generally available as line scanning devices, consisting of a single line of light detectors, and as area scanning devices, consisting of a rectangular 2D matrix of detectors; CMOS are commonly available as area devices. For Digital Holography only the latter architecture is of interest.

To illustrate the concepts of electronic sensors we will base our discussion around the CCD, but the principles of CMOS are broadly similar. Imaging using a CCD sensor is performed in a three-step process [26], involving,

1. Light exposure (the incident light on each pixel is converted into charges by the internal photo effect).
2. Charge transfer (the induced charge packets are moved through the semiconductor (silicon) substrate to memory/storage cells), and,
3. Charge to voltage conversion and output amplification (the capacitor matrix of the memory cells converts the transferred charge to a voltage; an amplifier adapts the voltage to the output requirements).

Three basic architectures are common in CCD sensors *viz.* interline transfer, frame transfer and full-frame transfer configurations respectively.

Fig. 3.16 Interline-transfer
architecture

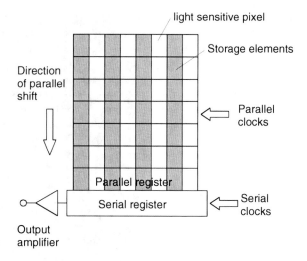

Interline (IL) transfer devices consist of rows of light-sensitive detector elements interleaved with rows of non-sensitive or light shielded storage elements, see Fig. 3.16. The charge packets which are generated in the light sensitive pixels are shifted into the adjacent storage area by a parallel clock; and are then shifted line-by-line into a serial register. The serial register transfers the charge packages to an amplified charge-to-voltage converter to form the output signal. The major disadvantage of interline transfer CCDs is their complexity, which results from separating the photo-detecting and storage (readout) functions.

Frame-transfer (FT) CCDs also have different areas for light conversion and for storage but are arranged into two area arrays rather than lines: a light sensitive capture area and a shielded storage area, see Fig. 3.17. The idea is to rapidly shift a captured scene from the photosensitive array to the storage array. The readout from the storage register is performed similarly to the readout process of interline transfer devices.

Full-Frame (FF) sensors have the simplest architecture, see Fig. 3.18. In contrast to IL and FT devices there is no separate storage area. The entire sensor area is light sensitive. The photons are converted into charge packets at each pixel and the resulting rows of image information are then shifted in parallel to the serial register, which subsequently shifts the row of information to the output as a serial stream of data. The process repeats until all rows are transferred off-chip. Since the parallel register is used for both image detection and readout, a mechanical shutter is needed to preserve scene integrity. Full-frame sensors have highest resolution and the production costs are comparably inexpensive.

In principle all three types of sensor are suitable for Digital Holography. Full frame type sensors have the advantage that the exposure time can be adjusted according to the demands of a specific application. Even exposure times in the range of seconds are possible. However, the mechanical shutter limits the number of holograms, which can be recorded per second (frame rate). In addition the shutter

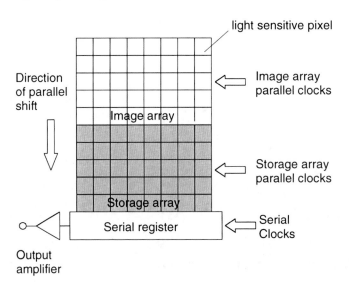

Fig. 3.17 Frame-transfer architecture

Fig. 3.18 Full-frame architecture

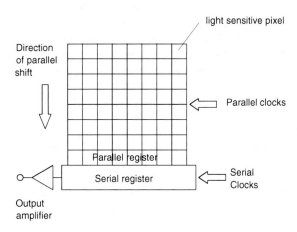

may cause mechanical vibrations to the set-up, which deteriorate the hologram quality. An advantage of interline transfer type sensors is that they are equipped with an electronic shutter, allowing higher frame rates. The best suited camera type depends therefore on the specific holographic application.

In contrast to CCDs each light sensitive pixel of a CMOS sensor is equipped with its own amplifier; i.e. the charge-to-voltage conversion is carried out at pixel level. Each pixel can be read out individually. State-of-the-art CMOS sensors have pixel pitches as small as 1.12 μm (see Table 3.1), which makes them an important alternative for digital holographic applications.

Table 3.1 CCD and CMOS cameras suitable for Digital Holography

Camera	Chip type	Number of pixels	Pixel size (μm^2)	Frames per second	Dynamic range	θ_{max} for $\lambda = 532$ nm
Roper Sci. MegaPlus 1.4i	CCD	1,317 × 1,035	6.8 × 6.8	6.9	8 bit	2.2°
GT3300	CCD	3,296 × 2,472	5.5 × 5.5		8 /14 bit	2.8°
Duncan DT1100	CCD	1,392 × 1,040	4.65 × 4.65	12	8 /10 bit	3.3°
DMK 72BUC02	CMOS	1,280 × 960	2.2 × 2.2	15	8 bit	6.9°
Sony CMX081PQ[a]	CMOS	No data	1.12 × 1.12	15	No data	13.6°

[a] For mobile phones

3.4.2 Spatial Frequency Requirements

The CCD or CMOS sensor records the interference pattern resulting from super-position of the reference wave with the waves scattered from the different object points. In order to reduce averaging effects over the area of a pixel, the maximum spatial frequency of the hologram should be smaller than the resolution limit imposed by the sensor. The maximum spatial frequency that can be resolved is determined by the maximum angle θ_{max} between the reference wave and the waves scattered from the different object points according to Eq. (2.30), and given by,

$$f_{max} = \frac{2}{\lambda} \sin \frac{\theta_{max}}{2} \qquad (3.51)$$

Photographic emulsions used in classical optical holography have resolutions up to 5,000 line pairs per millimetre (lp/mm). Using these materials, holograms with beam angles of up to 180° can be recorded. However, typical pixel dimensions of CCD/CMOS sensors are around $\Delta x \approx 5$ μm. Consequently, the corresponding maximum resolvable spatial frequency is given by

$$f_{max} = \frac{1}{2\Delta x} \qquad (3.52)$$

and is therefore in the range of 100 lp/mm for 5 μm pixels. Combining Eqs. (3.51) and (3.52) leads to a maximum angle, given by

$$\theta_{max} = 2 \arcsin \left(\frac{\lambda}{4\Delta x} \right) \approx \frac{\lambda}{2\Delta x} \qquad (3.53)$$

where the approximation is valid for small angles. For a recording wavelength of 532 nm and 5 μm pixels, the maximum recordable angle is about 3.0°. The pixel size therefore limits the maximum angle between the reference and object wave.

3.4.3 Cameras for Digital Hologram Recording

The principle parameters of some selected CCD and CMOS cameras suitable for Digital Holography are listed in Table 3.1.

The sensitivity of CCD or CMOS cameras is typically in the range of 10^{-4} to 10^{-3} J/m^2, which is higher than the sensitivity of photographic emulsions used for classical holography. The spectral response of silicon-based sensors covers the range from approximately 400–1,000 nm. Many commercial cameras are equipped with spectral filters to restrict the sensitivity to the visible spectrum.

In conventional holography with photographic plates the intensity ratio between reference and object wave is normally set to be in the range of 5:1–10:1 in order to avoid nonlinear effects due to the recording medium. However, the maximum contrast in an interference pattern is achieved if the intensity ratio between the two waves is 1:1. Electronic sensors have a much better linearity in the exposure curve than photographic emulsions and consequently, a unity intensity ratio is normally aimed for. As in classical holography the total light energy impinging on the sensor can be controlled by varying the exposure time using a mechanical or the electronic camera shutter.

Currently, CMOS cameras possess the highest resolution (smallest pixel size), see Table 3.1. On the other hand CMOS cameras often have a logarithmic exposure curve. However, this can be tolerated; the advantage of smallest pixel size is more important. The dynamic ranges of CCD- and CMOS-devices is typically 8 bit (256 grey values) or higher. This is comparable with photographic materials and fully sufficient for hologram recording. Even objects with brightness variations exceeding the dynamic range of the recording medium can be stored and reconstructed, because the object information is coded as interference pattern (hologram).

Efficient numerical reconstruction of digital holograms making use of the fast Cooley-Tukey FFT algorithm requires a pixel number, which is a power of 2 (e.g. 1,024 × 1,024). The pixel numbers of some of the cameras listed in Table 3.1 differ from that rule. For a pixel number of e.g. 1,317 × 1,035 (MegaPlus 1.4i) only 1,024 × 1,024 pixels are used for reconstruction. In the case of pixel number slightly lower than a power of 2 it is advisable to add artificial pixels with grey value zero (black) to the recorded hologram until a pixel number of $2^n \times 2^n$ is reached. This *zero padding* does not distort the reconstructed image; it only causes a smoothing or interpolation.

3.4.4 Recording Set-ups

In this section typical arrangements used in Digital Holography are discussed with respect to their spatial frequency limitations. In Fig. 3.19a a plane reference wave propagates perpendicularly to the sensor. The object is located off-axis with respect to the optic axis. This arrangement is very simple, but the space occupied by the object is not used efficiently. The maximum angle between rays emanating from the edge of a cubic object with sides of length L, to the opposite edge of the sensor with sides of length $N\Delta x$ is (distance x shown in Fig. 3.19) is given as

$$\theta_{\max} \approx \frac{x}{d_{\min}} = \frac{\sqrt{\frac{5}{4}}(L + N\Delta x)}{d_{\min}} \tag{3.54}$$

The corresponding minimum object distance $d_{\min}$ is calculated by equating this expression with the approximation for $\theta_{\max}$ in Eq. (3.53), and thus,

Fig. 3.19 Recording set-ups. *Left* side view; *Right* top view as seen from sensor

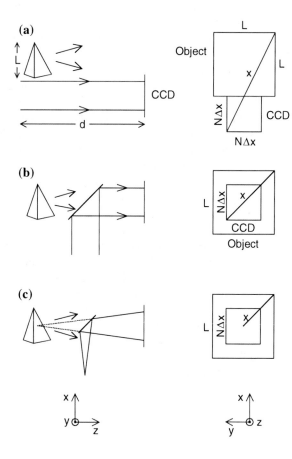

$$d_{min} = \sqrt{\frac{5}{4}\frac{2\Delta x}{\lambda}(L + N\Delta x)} = \sqrt{5}\frac{\Delta x}{\lambda}(L + N\Delta x) \qquad (3.55)$$

In Fig. 3.19b the plane reference wave is coupled into the set-up via a beam splitter. This allows positioning the object symmetrically, i.e. objects with larger dimensions can be recorded at a given distance d. The minimum object distance is:

$$d_{min} \approx \frac{x}{\theta_{max}} = \sqrt{2}\frac{\Delta x}{\lambda}(L + N\Delta x) \qquad (3.56)$$

However, the DC term is in the centre of the reconstructed image and has to be suppressed by the procedures described in Sect. 3.3.1.

Figure 3.19c shows an arrangement for lensless Fourier holography. The spherical reference wave is coupled into the set-up via a beam splitter in order to have the source point in the object plane. The minimum object distance is:

$$d_{min} = \sqrt{2}\frac{\Delta x}{\lambda}L \qquad (3.57)$$

In the lensless Fourier arrangement the shortest object distance can be chosen.

For all the arrangements shown, the maximum spatial frequency has to be adapted very carefully to the resolution of the sensor. If too high a spatial frequency occurs, the contrast of the entire hologram decreases or, in the extreme case, it vanishes totally. In practice, suitably placed apertures, which restrict the lateral propagation help to ensure that the spatial frequency requirements are met.

To record objects with dimensions larger than a few centimetres on a typical sensor, the recording distance d needs to be increased up to several meters. This may not be feasible in practice and recording arrangements are developed to maintain object angles within a resolvable spatial frequency spectrum [180, 203]. A typical example is shown in Fig. 3.20. A diverging lens is placed between the

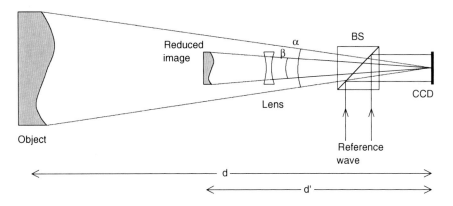

Fig. 3.20 Recording geometry for large objects

object and the target generates a de-magnified virtual image of the object at a distance d'. The wave field emerging from this virtual image is superimposed with the reference wave and the resulting hologram is recorded. The maximum spatial frequency is lower than that of a hologram recorded without object reduction.

3.4.5 Stability Requirements

A stable optical set-up is necessary for digital as well as for conventional holography. Any change in the optical path difference between the interfering beams will result in a movement of the fringes and reduced contrast in the hologram. In practice, the path variation should not exceed 1/4 to 1/10 of a wavelength during hologram exposure. For holography using a continuous wave laser it is essential to mount the optical arrangement on a vibration isolated table. For field holography, a short duration, of the order of a few nanoseconds, pulsed laser is a better option. In contrast to classical holography disturbances due to vibrations are visible in DH even in the recording process: the hologram visible on the monitor of the recording system has a low modulation or the contrast vanishes totally. This is an easy way to monitor the stability of the set-up against vibrations.

3.4.6 Light Sources

The coherence length L_c of the light source used for off-axis holography has to be longer than the optical path difference (OPD) between the reference and object wave paths (measured from the beam splitter to the recording medium) for recording of holograms. If L_c is too short the interference pattern between reference- and object wave vanishes. In practice for most applications the use of a laser is mandatory.

Some commonly used continuous (cw) lasers for Digital Holography and their typical specifications are summarized in Table 3.2. The most common lasers now

Table 3.2 Selected cw lasers for Digital Holography

Laser	Wavelength (nm)	Output power	Coherence length (m)
He–Ne-laser (multi-mode)	632.8	1–50 mW	0.2
Argon-ion laser (single mode)	488/514.5	Up to several W	Up to 100
Frequency doubled Nd:YAG-laser	532	Up to several W	Several 10's
Stabilized diode laser	Various	5–100 mW	Up to 100

used in DH are the frequency-double diode pumped solid state laser (FD-DPSS) and single mode diode lasers.

The FD-DPSS is usually, but not exclusively, based on a doped-insulator crystal such as Nd-YAG; its fundamental wavelength is 1,064 nm producing a frequency-doubled output of 532 nm (the green part of the visible spectrum) over coherence lengths of several tens of meters. It is available either in continuous wave mode (cw), with output powers up to several watts, or in pulsed mode with hundreds of millijoules output over a few nanoseconds duration and at pulse repetition rates of up to 50 Hz. Lower energy (tens of microjoules) models are available with repetition rates up to kilohertz. Flashlamp (rather than diode) pumping can produce energies of several joules. They are rapidly replacing gas and ruby lasers as the preferred option for classical and digital holography.

For field applications out with controlled laboratory conditions, or if moving objects have to be recorded, a pulsed laser is necessary. Formerly ruby lasers were commonly used. Now pulsed Nd:YAG-lasers have better characteristics with respect to compactness, pulse stability and repetition rate.

Diode lasers are also now commonly for continuous wave applications. Single mode operation can be achieved by stabilization electronics. Stabilized diode lasers have long coherence lengths and sufficient output power. However, the wavelength is not fixed by atomic transitions as for the lasers discussed above. That means it is necessary to monitor its wavelength during operation. In addition the wavelength depends on the temperature, typical drift is of the order of 0.2 nm/°C. On the other hand their wavelengths are tuneable over the order of several nanometers. Tuneable diode lasers are used for two-wavelength contouring for example.

Another interesting type of light source for DH is the superluminescent diode (SLED or SLD). Such diodes combine the high output power of laser diodes with the low temporal coherence of conventional LED's. These devices are the ideal choice, if low coherent noise but high brightness is necessary.

In the early days of holography, both classical and digital, gas lasers such as Helium-Neon (HeNe) and argon ion were almost exclusively used for continuous wave holography and ruby lasers for pulsed holography. The Helium-Neon (He–Ne) laser is able to operate at several different wavelengths, but mostly commonly the red 632.8 nm line is used. He–Ne lasers are moderately inexpensive, the technology is mature and still found in many laboratories and schools for educational uses of holography. When DH was in its infancy, and only sensors with pixel sizes of about 10 µm were available, the relatively long wavelength was advantageous, because it allowed larger angles between the interfering waves (see Sect. 3.4.2). Unstabilized He–Ne lasers oscillate on several longitudinal modes. The coherence length of such lasers is therefore not determined by the width of a single mode, but by the width of the entire gain profile and is in the order of 20 cm.

The spatial coherence of the laser is also crucial in holography. Only with an object illumination of sufficient spatial coherence it is possible to generate a scattered light field with defined complex amplitude in the far field domain of the object. The lasers discussed above usually oscillate in a single transverse mode (TEM$_{00}$) and have a Gaussian profile.

For some special applications in microscopy, particle sizing or in shearing interferometry the requirements on temporal coherence are lower than for off-axis holography. In this case light emitting diodes (LED's) can often be used. LED's have a spectral width of about 10 nm or, equivalently, a coherence length in the range of 50 μm. This is sufficient if the OPD is sufficiently low. In Chap. 7 we present some computational methods which enable sensing of low coherence wave fields as well. This enables applications very similar to those applicable for DH, such as numerical refocusing for example, but using low coherent light provided by an LED.

Chapter 4
Digital Holographic Interferometry (DHI)

4.1 General Principles

As we saw in Chap. 2, a conventional holographic interferogram recorded on photographic film is generated by superposition of two waves, which are scattered from an object in two different states of loading or excitation. The interferogram carries the information about the phase change between the two waves in the form of dark and bright lines, or interference fringes. However, the interference phase cannot be extracted unambiguously from a single interferogram; it is usually calculated from three or more phase shifted interferograms by phase shifting algorithms. This requires additional experimental and processing effort.

Digital Holography allows a completely different way of processing [195]. In each state of the object one digital hologram is recorded. Instead of superimposing these holograms as in classical HI, the digital holograms are reconstructed separately according to the theory presented in Chap. 3. From the resulting complex amplitudes $\Gamma_1(\xi, \eta)$ and $\Gamma_2(\xi, \eta)$ the phases are obtained:

$$\varphi_1(\xi, \eta) = \arctan \frac{\mathrm{Im}\,\Gamma_1(\xi, \eta)}{\mathrm{Re}\,\Gamma_1(\xi, \eta)} \tag{4.1}$$

$$\varphi_2(\xi, \eta) = \arctan \frac{\mathrm{Im}\,\Gamma_2(\xi, \eta)}{\mathrm{Re}\,\Gamma_2(\xi, \eta)} \tag{4.2}$$

The superscripts 1 and 2 denote the first and second states of excitation, respectively. In Eqs. (4.1) and (4.2) the phase takes values between $-\pi$ and π, the principal values of the arctan function. The interference phase is now calculated directly by subtraction:

$$\Delta\varphi = \begin{cases} \varphi_1 - \varphi_2 & \text{if } \varphi_1 \geq \varphi_2 \\ \varphi_1 - \varphi_2 + 2\pi & \text{if } \varphi_1 < \varphi_2 \end{cases} \tag{4.3}$$

© Springer-Verlag Berlin Heidelberg 2015
U. Schnars et al., *Digital Holography and Wavefront Sensing*,
DOI 10.1007/978-3-662-44693-5_4

This equation permits the calculation of the interference phase modulo 2π directly from the digital holograms. The generation and evaluation of an interferogram is not necessary.

The Digital HI process is shown in Fig. 4.1. The upper left and upper right figures present two digital holograms, recorded in different states. Between the two recordings, the knight has been tilted by a small amount. Each hologram is reconstructed separately by a numerical Fresnel transform. The reconstructed phases according to Eqs. (4.1) and (4.2) are depicted in the two figures of the middle row. The phases vary randomly due to the surface roughness of the object. Subtraction of the phases according to Eq. (4.3) results in the interference phase, lower left figure.

The interference phase is indefinite to an additive multiple of 2π, i.e. the wrapped modulo 2π information about the additive constant is already lost in the holographic interferometric process. This property is not exclusive to Digital HI, but is also the case for all interferometric methods using the wavelength as a length unit. To convert the interference phase modulo 2π into a continuous phase distribution, one can apply the standard phase unwrapping algorithms developed for conventional interferometry, HI or ESPI. In this example a simple path dependent unwrapping algorithm, as described in Sect. 2.7.6 has been applied. The unwrapped phase image is shown in the lower right picture of Fig. 4.1. The sensitivity vector used for hologram recording is nearly constant and perpendicular over the whole surface. The grey values of the unwrapped phase map can be converted therefore directly into displacement values via Eq. (2.86), i.e. the plot in the lower right picture of Fig. 4.1 is the object displacement.

4.2 Deformation Measurement

4.2.1 Quantitative Displacement Measurement

As discussed in Sect. 4.1 the way to obtain the interference phase in DHI is totally different from conventional HI using photographic recording media and optical reconstruction. On the other hand, the derivation of the relationship between the displacement vector $\vec{d}$, the geometrical boundary conditions described by the sensitivity vector $\vec{S}$, and the interference phase $\Delta\varphi$ is also valid for DHI. That means the deformation is calculated by Eq. (2.84), which is repeated here:

$$\Delta\varphi(x, y) = \frac{2\pi}{\lambda}\vec{d}(x, y, z)\left(\vec{b} - \vec{s}\right) = \vec{d}(x, y, z)\vec{S} \qquad (4.4)$$

As an example of quantitative displacement measurement, the deformation of a plate due to impact loading is discussed [201, 203]. The plate is made of fibre reinforced plastic (FRP), which is used more and more in aircraft industry. The

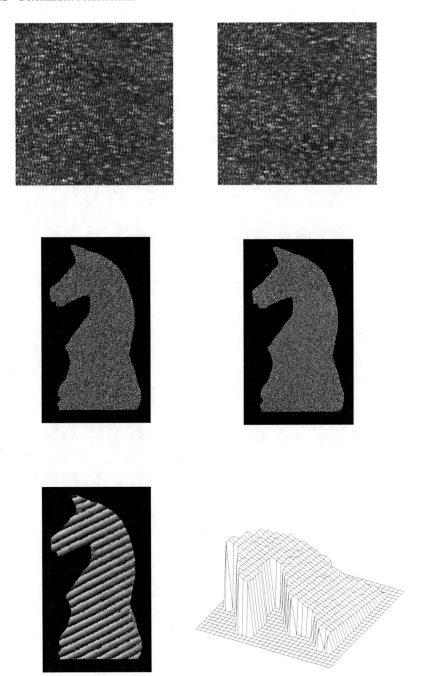

Fig. 4.1 Digital holographic interferometry

deformation behaviour of FRP under impact loading differs from that of static loading, so impact experiments are necessary. DHI is well suited to measurement of such transient deformations, because only one single recording is necessary in each deformation state.

The holographic set-up is shown in Fig. 4.2. The dimensions of the plate are 12 cm × 18 cm. The recording distance would be too long for direct recording. The spatial frequency spectrum is therefore reduced by a lens, as explained in Sect. 3.4.4 (set-up in Fig. 3.20).

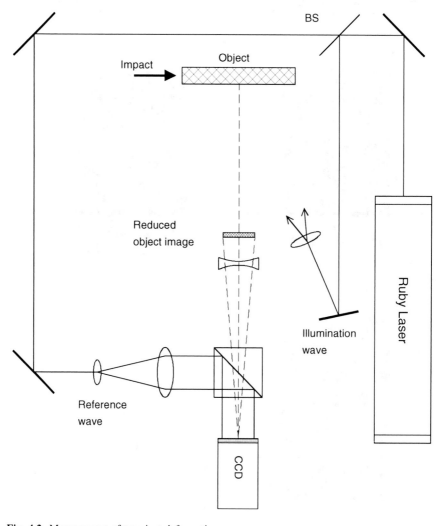

Fig. 4.2 Measurement of transient deformations

The plate is clamped at three sides by a stable frame. A pneumatically accelerated steel projectile hits the plate and causes a transient deformation. Two holograms are recorded: The first exposure takes place just before impact, when the plate is in rest. The second hologram is recorded 5 µs after the impact. The holograms are recorded by a pulsed ruby laser with a pulse duration of about 30 ns, short enough for hologram recording of dynamic events. Recording of the second hologram is triggered by a photoelectric barrier, which generates the start signal for the laser after the projectile has crossed. The time interval between the impact and the second laser pulse is adjustable by an electronic delay.

Both holograms are reconstructed separately as described in Chap. 3. The interference phase map is then calculated by subtracting the reconstructed phase distributions according to Eq. (4.3).

As a typical result, the interference phase modulo 2π and the unwrapped phase are shown in Figs. 4.3 and 4.4. Since the sensitivity vector is nearly perpendicular to the surface, the unwrapped phase corresponds to the deformation field in z-direction 5 µs after impact.

Fig. 4.3 Interference phase modulo 2π

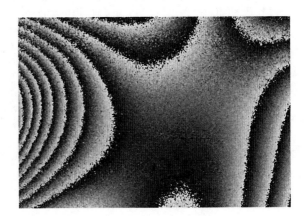

Fig. 4.4 Unwrapped phase, corresponding to deformation 5 µs after the impact

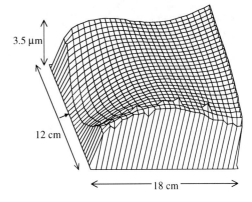

4.2.2 Mechanical Materials Properties

Digital holographic displacement measurement can be used to determine mechanical and thermal material properties such as, Young's modulus, the Poisson ratio and the thermal expansion coefficient [101, 208]. For the derivation of these quantities a procedure for the evaluation of the three-dimensional shape and the three-dimensional displacement of the object under test, a physical model of the behavior of the loaded object and the knowledge about the applied load are necessary. The following description of such a DHI measurement procedure is partly based on [208].

The physical model must contain one or more of the material constants as parameters. A numerical fit into the measured data according to the physical model delivers the wanted parameters within an accuracy that is determined by the numerical reliability of the model. An outline of the complete evaluation process is shown in Fig. 4.5 (DHI shape measurement is explained in Sect. 4.3).

The calculation of the above mentioned material quantities requires measurement of the whole three-dimensional displacement vector field. As indicated in Sect. 2.7, at least 3 interferograms of the same surface with linear independent sensitivity vectors are necessary. The interferometer consists of an optimized arrangement with 4 illumination directions and 1 observation direction to precisely measure the 3D-displacements and coordinates, as in Fig. 4.6. The interferometer incorporates a CCD-camera, a laser, a beam splitter cube to guide the reference beam to the CCD target and beam shaping optics. Optionally, a fibre coupler can be included to switch several illumination directions for varying the sensitivity vector. Such an interferometer can be very compact in its design, Fig. 4.6. Small silicon beams are used as test samples, Fig. 4.7.

Here we describe the use of the above DHI system to determine the Poisson ratio of a given material. Figure 4.8a shows a typical loading machine designed especially for small objects. The small dimensions of the samples demand a precise adjustment of all components, including the bolt which pushes against the object from above, the support and the sample which has the shape of a small rectangular

Fig. 4.5 Flowchart for evaluation of material properties using DHI (from [208])

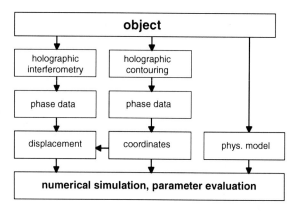

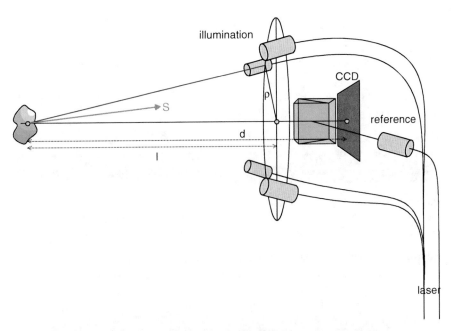

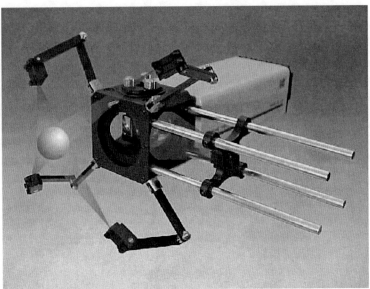

Fig. 4.6 DHI set-up with four illumination directions (*top*) and its practical implementation (*bottom*, *photo* BIAS)

beam. In this way a homogeneous deformation is achieved. Unwanted torsions of small magnitude are corrected numerically. This can be done easily with the use of the modulo 2π-phase maps from Digital Holography. The resulting deformation is

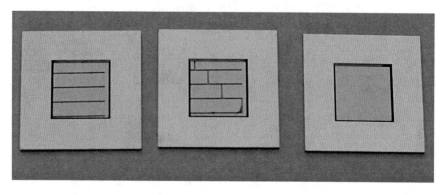

Fig. 4.7 Test samples made on 100 mm diameter silicon wafer. The size of the quadratic structure is 9 mm × 9 mm, the thickness of the components is between 10 and 40 μm (*photos* BIAS)

recorded and evaluated. The deflection produces a hyperbolic pattern in the 2π-phase map, Fig. 4.8b. Conventionally, the Poisson ratio is derived numerically from the angle between the asymptotic lines of the fringes of equal phase [254]. The deformation can be formulated by the following equation to a first order approximation:

$$u(y, z) = -\frac{1}{2R}\left[z^2 + v\left(a^2 - y^2\right)\right] \tag{4.5}$$

where u describes the deformation in the x-direction at a position (y, z) on the surface of the object, v is the Poisson ratio, R is the radius of curvature and a is a constant parameter. Equation (4.5) shows that the upper and lower surface of the sample are deformed to parabolic curves where the inside is bent in a convex profile and the outside is concave. Since this analytical model contains the Poisson ratio as a parameter it is possible to use the measured deformation for its evaluation. This is performed numerically by approximating the model to the data (Fig. 4.8c) with a least-square-fit, Fig. 4.8d.

The reproducibility and accuracy of the values obtained by this method is good in comparison to conventional optical techniques for small samples. Table 4.1 contains some of the results for beams made of spring steel, structural steel and titanium. The values correlate with the values given by the manufacturers within the tolerances of the material batches.

Young's modulus can be determined in a similar way to the Poisson ratio if the physical model contains this quantity as a parameter. Small silicon beams are clamped at one edge and mechanically loaded at the opposite edge with a defined force. The 3D-surface displacement (u,v,w) (Fig. 4.9c) can be measured with the interferometer by evaluating at least 3 interferograms (Fig. 4.9b) made with different illumination directions. A model of the beam bending containing the Young's modulus E as a free parameter is the basis for a numerical fit of the experimental values:

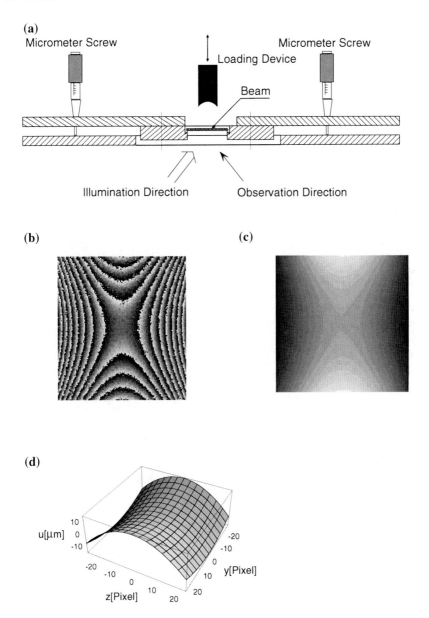

Fig. 4.8 Measurement of the poisson ratio by DHI (from [208]). **a** Schematic experimental set-up. **b** Reconstructed mod 2π-phase map. **c** Unwrapped phase field. **d** Approximated function

Table 4.1 Measured poisson ratios compared with literature values (after [208])

Material	Width (mm)	Thickness (mm)	Length (mm)	Poisson ratio measured	Poisson ratio literature
Spring steel	1.20	0.20	12.0	0.288	0.29–0.31
Spring steel	2.00	0.10	12.0	0.301	0.29–0.31
Structural steel	1.00	0.50	10.0	0.338	0.29–0.31
Structural steel	1.50	0.50	10.0	0.345	0.29–0.31
Titanium	2.00	0.80	10.0	0.359	0.361
Titanium	1.00	0.80	10.0	0.381	0.361

$$u(y) = \frac{Fl^3}{6EI_y}\left(2 - 3\frac{y}{l} + \frac{y^3}{l^3}\right) \tag{4.6}$$

where u is the displacement in x-direction and y is a position on the beam of the length l. I_y is the axial moment of inertia in the (x,z)-plane that can be estimated with the help of a shape measurement and F is the force applied to the tip of the beam. The applied forces are relatively small so that a special loading mechanism was developed, Fig. 4.9a. The spring constant k is assumed to be known precisely as is the displacement $\Delta a = a - a'$. With this information the force can be evaluated from

$$F = k\Delta a \tag{4.7}$$

The experiments with thin beams made of silicon (dimensions: length 3 mm, width 1 mm) delivered an average value of E = 162 MPa. The literature value (in the considered crystal direction) is typically about 166 MPa, but can vary widely depending to the material's history, treatment and degree of impurity.

4.2.3 Thermal Materials Properties

DHI is applied also to measure of the thermal properties of a material, e.g. its thermal expansion coefficient [101, 208]. For interferometric investigations of thermal behavior thermal turbulence and non-uniform temperature distributions should be avoided. Therefore a vacuum chamber is used that can be supplied with adapted loading devices, Fig. 4.10a. The thermal loading device is capable of keeping a constant temperature within an accuracy of 0.02 °C in a range of about 20 °C up to 180 °C, Fig. 4.10b. The digital holographic interferometer is mounted outside at the observation window of the chamber, Fig. 4.10c.

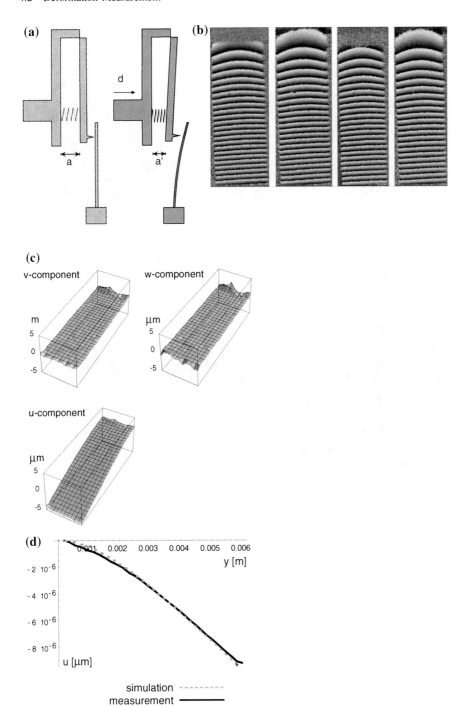

Fig. 4.9 Determination of Young's modulus by DHI (from [208]). **a** Working principle of the loading mechanism for the small samples. **b** four 2-phase maps recorded from four different illumination directions. **c** Deformation calculated in cartesian coordinates (scale of the plots in µm). **d** Profile of the deformation in x-direction

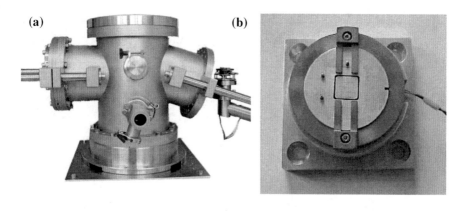

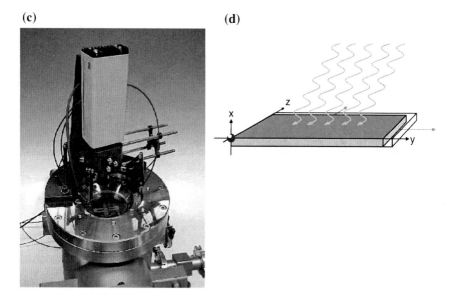

Fig. 4.10 Determination of the thermal expansion coefficient by DHI (*photos* BIAS). **a** Vacuum chamber with the supply channel. **b** Equipment for thermal loading. **c** Interferometer mounted on the inspection window. **d** Coordinate system used for the calculation of the thermal expansion coefficient

A mono-crystal silicon beam (Fig. 4.10d) with a phosphor coating is used as a test object. The interferograms are recorded at various temperature differences. The complete evaluation process can be summarized as follows:

- 4 holograms are recorded with the object in its initial state
- the object is loaded thermally and recorded holographically from four different illumination directions

- the displacement vector components *(u,v,w)* are calculated based on the evaluation of the four 2π-phase maps
- rigid body motions are separated from internal deformations of the object itself by subtracting the mean movement from the displacement values.
- the absolute length change ΔL is determined as well as the total length of the beam which can be performed by using the imaging properties of Digital Holography.
- The thermal expansion coefficient in y- and z-direction can simply be calculated by using the equation

$$\alpha = \frac{\Delta L}{L_0 \Delta T} \tag{4.8}$$

However, the extension in the x-direction is too small to be detected with this method.

As an example, the thermal expansion coefficient α of a 2 mm × 9 mm × 100 μm mono-crystal silicon beam was measured. Figure 4.11 shows the four resulting 2π-phase maps. The applied temperature difference ΔT is 30 °C. After elimination of the rigid body motion the three deformation components are evaluated as shown in Fig. 4.12. When the beam dimensions are taken into account a value of about $\alpha = 2.92 \times 10^{-6}$ $1/K$ is obtained. Literature values vary over a wide range due to different measurement methods, conditions and material batches: $\alpha = 2.4–6.7 \times 10^{-6}$ $1/K$.

4.2.4 Non-destructive Testing

Non-Destructive Testing (NDT) is a generic term used to describe any method of measuring or testing materials, components and structures without damaging them in any way. Some of the most common NDT techniques used in industry include,

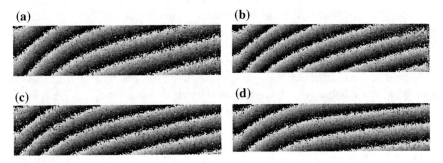

Fig. 4.11 2π-phase maps due to deformation by thermal loading, four different illumination directions. **a** Illumination direction 1. **b** Illumination direction 2. **c** Illumination direction 3. **d** Illumination direction 4

Fig. 4.12 3D-displacement
vector components (u,v,w) of
thermally loaded object. **a** u-
component. **b** v-component.
c w-component

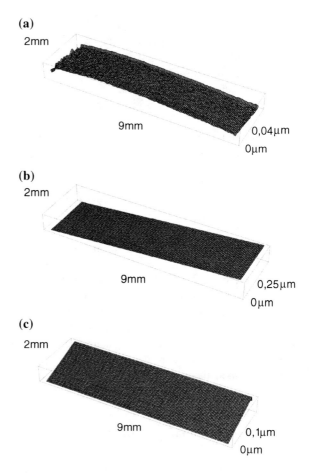

ultrasonics, eddy currents, dye-penetrants testing, X-rays and, importantly in this
context optical methods like HI, ESPI and shearography.

Holographic Non-Destructive Testing (HNDT) measures the deformation due to
mechanical or thermal loading of a specimen. Flaws inside the material create a
surface deformation which is detected as an inhomogeneity in the holographic
fringe pattern.

HNDT can be used wherever the presence of a structural weakness results in a
characteristic surface deformation of the stressed component. The load can be
realized by the application of a mechanical force or by a change in pressure or
temperature. Holographic NDT indicates deformations down to the submicrometer
range, so loading amplitudes far below any damage threshold are sufficient to
produce detectable fringe patterns.

In HNDT it is sufficient to have one fringe pattern of the surface under investigation. Quantitative evaluation of the displacement vector field is usually not
required. The fringe pattern is evaluated qualitatively by an observer or, more

frequently, by fringe analysis computer codes. Irregularities in the interference pattern are indicators of flaws within the component under investigation.

As discussed in the preceding chapters, DHI does not generate a fringe pattern, but instead directly produces an interference phase map from which any flaws can be determined.

To illustrate the processes of DHI NDT, we describe testing of a pressure vessel [202] as those used as gas tanks in satellites, see Fig. 4.13. The diameter of the vessel is in the order of 1 or 2 m and the thickness of the wall is only about 1 mm. Typical flaws to be detected are cracks or reduced thickness of the wall.

The surface of the tank is divided into segments of about 5 cm × 5 cm. For each segment, a series of digital holograms is recorded. Between the exposures, the pressure inside the tank is varied by a few hundred hPa. As a typical result the interference phase between one pair of holograms is shown in Fig. 4.14. The disturbance in the middle is an indication of a flaw. The interference phase can also be converted into a continuous phase by unwrapping the 2π-jumps. However, for flaw detection the unwrapped phase map is often more suitable.

Fig. 4.13 Satellite tank

Fig. 4.14 Non-destructive
testing by DHI

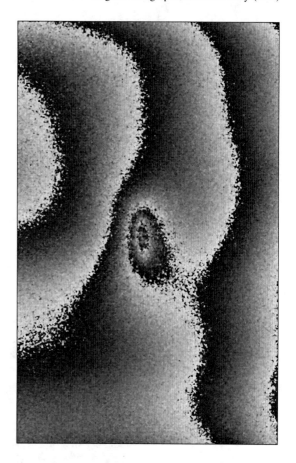

In conventional holographic NDT the load applied to the object under investigation is restricted to a certain limit. Loads above this limit produce such high fringe densities that they are unresolvable. In numerical reconstruction in DHI, the phase difference between any pair of exposures can be calculated. Even if the total deformation between the first and the last hologram is too large for direct evaluation, the total phase difference can be calculated stepwise as the sum of the individual phase changes:

$$\Delta\varphi_{total} = \Delta\varphi_{1\rightarrow2} + \Delta\varphi_{2\rightarrow3} + \Delta\varphi_{3\rightarrow4} + \cdots \tag{4.9}$$

However, a drawback of DHI compared to conventional HI should be emphasized: For visual flaw detection it is sometimes an advantage to continuously vary (dynamic evaluation) the observation direction. This is possible for holograms recorded on a photographic plate with dimensions of about 10 cm × 10 cm or more, but until now not possible for digital holograms recorded on sensors with typically only about 1 cm × 1 cm area. However, future progress in camera technology and computer hardware may solve this problem.

4.3 Shape Measurement

4.3.1 Two-Illumination-Point Method

The two contouring techniques discussed in Sect. 2.7.3 for conventional HI are also applied in DHI.

For the Two-Illumination-Point method it is necessary to record two digital holograms of the same surface, but from different object illumination points. Both holograms are reconstructed separately. The interference phase map, which represents the object shape, is then calculated by subtracting the individual phase distributions according to Eq. (4.3). The result is a wrapped phase map, which is interpreted similar to the contour fringe pattern discussed in Sect. 2.7.3. The phase change between two adjacent 2π-jumps is

$$\Delta\varphi = \frac{2\pi}{\lambda}\vec{p}\vec{s} \qquad (4.10)$$

with $\vec{p}$ and $\vec{s}$ as defined previously. By analogy with Eq. (2.96) the distance between two adjacent 2π-jumps is

$$\Delta H = \frac{\lambda}{2\sin\frac{\theta}{2}} \qquad (4.11)$$

where θ is the angle between the two illumination directions.

DHI two-illumination-point contouring can be carried out with for example the set-up depicted in Fig. 4.15. Optical fibres are preferably used to guide the illumination wave. The output at fibre face is the illumination source point. The shift is realized by e.g. a motor driven translation stage. The first digital hologram is recorded with the fibre end at position S. For the second hologram the fibre is shifted slightly to position S'.

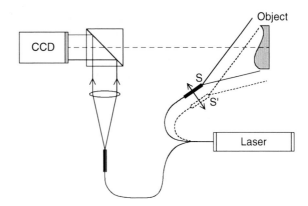

Fig. 4.15 Two-illumination point DHI

In order to obtain maximum sensitivity in a direction normal to the surface, illumination should come from the side, i.e. the angle between illumination direction and observation direction is near 90°. Yet, such a flat illumination may cause shadows due to surface variations. The optimum illumination direction is therefore always a compromise between maximum sensitivity and minimum shadows in the reconstructed images.

4.3.2 Two- and Multi-wavelength Method

For shape measurement by the two-wavelength method two holograms are recorded with different wavelengths λ_1 and λ_2. In conventional HI both holograms are recorded on a single photographic plate. Both holograms are reconstructed by the same wavelength, e.g. λ_1. That is why two images of the object are generated. The image recorded and reconstructed by λ_1 is an exact duplicate of the original object surface. The image which has been recorded with λ_2, but reconstructed with λ_1 is slightly shifted in towards the observer, (see imaging equations in Sect. 2.6.2) with respect to the original surface. The two reconstructed images interfere.

The concept of two-wavelength contouring has also been introduced into Digital Holography [103, 196, 207]. Two holograms are recorded with λ_1 and λ_2 and stored electronically, e.g. with the set-up depicted in Fig. 4.16. In contrast to conventional HI using photographic plates, both holograms can be reconstructed separately by the correct wavelengths according to the theory of Chap. 3. From the resulting complex amplitudes $\Gamma_{\lambda_1}(\xi,\eta)$ and $\Gamma_{\lambda_2}(\xi,\eta)$ the phases are calculated:

$$\varphi_{\lambda 1}(\xi,\eta) = \arctan\frac{\mathrm{Im}\Gamma_{\lambda 1}(\xi,\eta)}{\mathrm{Re}\Gamma_{\lambda 1}(\xi,\eta)} \qquad (4.12)$$

Fig. 4.16 Two-wavelength DHI

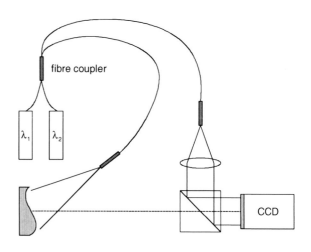

$$\varphi_{\lambda 2}(\xi, \eta) = \arctan \frac{\text{Im}\Gamma_{\lambda 2}(\xi, \eta)}{\text{Re}\Gamma_{\lambda 2}(\xi, \eta)} \tag{4.13}$$

As in deformation analysis the phase difference is now calculated directly by subtraction:

$$\Delta\varphi = \begin{cases} \varphi_{\lambda 1} - \varphi_{\lambda 2} & \text{if } \varphi_{\lambda 1} \geq \varphi_{\lambda 2} \\ \varphi_{\lambda 1} - \varphi_{\lambda 2} + 2\pi & \text{if } \varphi_{\lambda 1} < \varphi_{\lambda 2} \end{cases} \tag{4.14}$$

This phase map is equivalent to the phase distribution of a hologram recorded with the synthetic wavelength Λ. In conventional two-wavelength contouring the distance between adjacent fringes corresponds to a height step of $\Lambda/2$, see Eq. (2.90). Similarly in two-wavelength DHI a 2π phase jump corresponds to a height step of $\Lambda/2$:

$$\Delta H = \frac{\lambda_1 \lambda_2}{2|\lambda_1 - \lambda_2|} = \frac{\Lambda}{2} \tag{4.15}$$

A typical example of two-wavelength contouring is shown in Fig. 4.17.

In DHI contouring both holograms are reconstructed with the correct wavelength. Distortions resulting from hologram reconstruction with a different wavelength from the recording wavelength, as in conventional HI contouring, are therefore avoided.

A modified contouring approach, which is referred to as *multiwavelength contouring*, needs more than two illumination wavelengths to eliminate ambiguities inherent to modulo 2π phase distributions [111, 112, 172, 246]. The advantage of this technique is that it can also be used with objects that have phase steps or isolated object areas.

An example of a set-up for shape measurement of small objects is shown in Fig. 4.18 [52]. Contouring of small objects is useful e.g. in manufacturing control of electronic components or to monitor the growth of living biological samples.

A long-distance microscope objective (LDM) with a magnification of $M = 10$ and a numerical aperture of $NA = 0.28$ is used to image the sample. The working

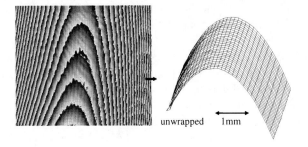

Fig. 4.17 Shape registration of a cylindrical lens by two-wavelength contouring. Visible part: 3 mm × 2 mm

Fig. 4.18 Holographic
microscope for shape
measurement of small objects

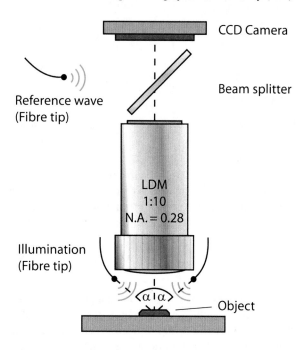

distance is 33 mm, so that the optical fibre for object illumination can be mounted
between the LDM and the sample. The CCD-sensor has 2,048 by 2,048 pixels with
a pixel pitch of 3.45 μm and is placed in a distance d behind the image. This offers
enough space to insert the beam splitter, which is necessary to superimpose the
reference wave. The CCD records therefore a hologram generated by a defocussed
image and the reference wave. The object image can be reconstructed with the
convolution approach as described in Sect. 3.2.2.

A tunable dye-laser is used as light source. For contouring, two holograms are
recorded at two wavelengths $\lambda_1 = 580$ *nm* and $\lambda_2 = 583$ nm. The synthetic wavelength according to Eq. (2.90) is therefore $\Lambda = 112$ μm. The holograms are
reconstructed and a phase image with contour lines is calculated. The object illumination angle chosen here is $\alpha = 20°$ with respect to the optical axis. With this
angle α, it is possible to calculate a quantitative height profile from the phase map.
Please note: the derivation in Sect. 2.7.3 is based on perpendicular illumination
($\alpha = 0°$), but the theory can be generalized to arbitrary angles.

As an example the surface profile of a small detail from a 2 € coin is measured,
see Fig. 4.19. The phase distribution from the marked detail in Fig. 4.19a is shown
in Fig. 4.19b. A principal drawback of two-wavelength holographic contouring is
the speckle decorrelation noise due to the different wavelengths. This leads to an
uncertainty in the phase determination. This effect is visible in the marked rectangle
of Fig. 4.19b: The phase information is nearly lost. A reduction of noise due to
speckle decorrelation can be achieved by averaging the phase maps from several
holograms, which are recorded with different illumination directions. This is

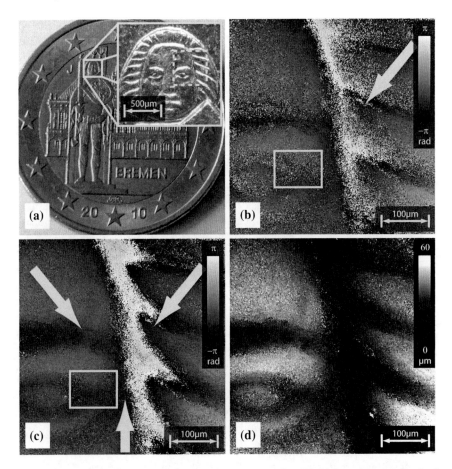

Fig. 4.19 Holographic microscope for shape measurement of small objects. **a** Photograph of a 2 Euro coin. **b** Phase distribution, calculated from a hologram with single illumination direction. **c** Average phase distribution obtained from three illumination directions. **d** Surface profile after unwrapping. The *arrows* indicate the projected directions of illumination

indicated in Fig. 4.19c, where the average phase distribution obtained from three illumination directions provides information even in the area marked by the rectangle. The surface profile of the coin after unwrapping is shown in Fig. 4.19d.

4.3.3 Hierarchical Phase Unwrapping

As stated earlier the process of phase unwrapping is always the same for conventional HI as well as for DHI and in general also for all methods which generate modulo 2π images. In Sect. 2.7.6 a simple phase unwrapping method is described.

However, for DHI multiwavelength contouring a special unwrapping procedure named *hierarchical phase unwrapping* is useful. The basic idea of this technique was originally developed by Nadeborn and Osten for projected fringe techniques with incoherent light [159, 169], but it is applicable for all measurement techniques which generate periodic data. *Hierarchical phase unwrapping* is particularly well suited in conjunction with DHI multi-wavelength contouring. The technique is discussed in this context.

The practical application of interferometric contouring techniques can lead to the following problems [206]:

- Fringe counting problem: The interference phases are periodic. For the Two-Wavelength method the periodic length is half the synthetic wavelength, see Eq. (4.15). If edges, holes, jumps or other discontinuations are on the measured surface, it is often not possible to count the absolute interference order or phase value. An unambiguous evaluation is not possible. In order to generate unambiguous results, it is therefore necessary to use a synthetic wavelength greater than twice the maximum height variation of the object. But this causes the second problem:

- Noise problem: The synthetic wavelength has to be adapted to the largest height jump in order to generate unambiguous phase values over the entire surface. The measurement of smaller height profiles with the same synthetic wavelength leads to larger phase noise compared to a wavelength well adapted to the profile. To measure the surface profile of an object with about 10 mm height variations, a synthetic wavelength of at least $\Lambda = 20$ mm is necessary. In practice the phase noise limits the achievable measurement resolution to about 1/10 of the wavelength; i.e. the accuracy of the measured height data is only 2 mm. This is too low if smaller height jumps of only 1 mm are measured.

The basic idea of hierarchical phase unwrapping is to start with a large synthetic wavelength to avoid phase ambiguities [246]. This measurement is not very accurate due to noise. The accuracy is now improved by systematic reduction of the synthetic wavelengths, while the information of the preceding measurements is used to eliminate ambiguities.

The procedure starts with a synthetic wavelength Λ_1, which is larger than twice the maximum height variation of the object. The height at a certain position is then given by

$$z_1 = \frac{\Lambda_1}{2} \frac{\Delta\varphi_1}{2\pi} \qquad (4.16)$$

where $\Delta\varphi_1$ is the measured interference phase at this position. This result is unambiguous, but has low accuracy. Now the synthetic wavelength is reduced to Λ_2 and a new phase measurement is made. The resulting height coordinate

$$\hat{z}_2 = \frac{\Lambda_2}{2} \frac{\Delta\varphi_2}{2\pi} \qquad (4.17)$$

is not unambiguous, this is indicated by the "^". In the next step the difference between the result of the first measurement z_1 and the second measurement $\hat{z}_2$ is calculated:

$$\Delta z = z_1 - \hat{z}_2 \qquad (4.18)$$

Furthermore which multiple of the periodic length $\Lambda_2/2$ is contained in the difference Δz (rounded number) is also calculated

$$N = floor\left(2\frac{\Delta z}{\Lambda_2} + 0.5\right) \qquad (4.19)$$

The function $f(x) = floor(x)$ delivers the maximum integer value, which is smaller than x. The correct result of the second measurement is now:

$$z_2 = \hat{z}_2 + \frac{\Lambda_2}{2} N \qquad (4.20)$$

This result is unambiguous as well as z_1, but it has a better accuracy compared to z_1 due to the smaller synthetic wavelength. The procedure is continued with smaller wavelengths as long as the resolution is below a desired value. After n iterations we arrive at:

$$z_n = \hat{z}_n + \frac{\Lambda_n}{2} floor\left(2\frac{z_{n-1} - \hat{z}_n}{\Lambda_n} + 0.5\right) \qquad (4.21)$$

In practice the number of measurements to reach the desired resolution should be as small as possible. This minimum or optimum number depends on the noise. Let ε_n be the inaccuracy in a measurement using the synthetic wavelength Λ_n. The true height coordinate z_{true} lies within an interval limited by

$$z_{max} = z_{meas} + \frac{\varepsilon_n}{2} \frac{\Lambda_n}{2}$$
$$\qquad (4.22)$$
$$z_{min} = z_{meas} - \frac{\varepsilon_n}{2} \frac{\Lambda_n}{2}$$

where z_{meas} is the value determined by the measurement. Then the next measurement with Λ_{n+1} takes place. The inaccuracy of this measurement is given by ε_{n+1}. Now the noise of the nth and the (n + 1)th measurement is considered for estimating the interval limits of the true height coordinate:

$$z_{max} = z_{meas} + \frac{\varepsilon_n}{2}\frac{\Lambda_n}{2} + \frac{\varepsilon_{n+1}}{2}\frac{\Lambda_{n+1}}{2}$$
$$z_{min} = z_{meas} - \frac{\varepsilon_n}{2}\frac{\Lambda_n}{2} - \frac{\varepsilon_{n+1}}{2}\frac{\Lambda_{n+1}}{2}$$

(4.23)

The difference between the maximum possible and the minimum possible height coordinate is:

$$|z_{max} - z_{min}| = \varepsilon_n \frac{\Lambda_n}{2} + \varepsilon_{n+1}\frac{\Lambda_{n+1}}{2}$$

(4.24)

A correct recovery of the absolute height coordinate is only possible, if following condition is satisfied for the (n + 1)th measurement:

$$\frac{\Lambda_{n+1}}{2} \geq |z_{max} - z_{min}|$$

(4.25)

A smaller value for $\Lambda_{n+1}/2$ than $|z_{max} - z_{min}|$ would lead to ambiguities, because $\hat{z}_{n+1} + N\Lambda_{n+1}/2$ as well as $\hat{z}_{n+1} + (N + 1)\Lambda_{n+1}/2$ are possible height values within the interval limits. The optimum period (half synthetic wavelength) is achieved for the equals sign.

The combination of Eqs. (4.24) and (4.25) with the equals sign results to:

$$\Lambda_{n+1} = \Lambda_n \frac{\varepsilon_n}{1 - \varepsilon_{n+1}}$$

(4.26)

This condition determines the optimum choice of the sequence of synthetic wavelengths depending on each measurement's accuracy.

4.4 Measurement of Refractive Index Variations

Digital HI is also used to measure refractive index variations within transparent media, e.g. with the set-up of Fig. 4.20. The expanded laser beam is divided into reference and object beam. The object beam passes the transparent phase object and illuminates the CCD. The reference beam impinges directly on the CCD. Both beams interfere and the hologram is digitally recorded. The set-up is very similar to a conventional Mach-Zehnder interferometer. The difference is that the interference figure here is interpreted as a hologram, which can be reconstructed with the theory of Chap. 3. Therefore all features of Digital Holography like direct access to the phase or numerical re-focussing are available.

Just like for deformation analysis two digital holograms are recorded: The first exposure takes place before, and the second after the refractive index change. These digital holograms are reconstructed numerically. From the resulting complex

Fig. 4.20 DHI set-up for transparent phase objects

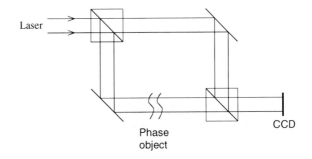

amplitudes $\Gamma_1(\xi, \eta)$ and $\Gamma_2(\xi, \eta)$ the phases are calculated by Eqs. (4.1) and (4.2). Finally the interference phase is calculated by subtraction according to Eq. (4.3).

In the reconstruction of holograms recorded by the set-up of Fig. 4.20 the non-diffracted reference wave, the real image and the virtual image are lie on the same axis. The images overlap, which causes distortion. The non-diffracted reference wave can be suppressed by filtering with the methods discussed in Sect. 3.3.1. The overlapping of the unwanted twin image (either the virtual image if one focuses on the real image or vice versa) can be avoided by slightly tilting the reference wave, as discussed in Sect. 3.3.2. In this case the images are spatially separated.

The interferometer of Fig. 4.20 is sensitive to local disturbances due to imperfections in optical components or dust particles. The influence of these disturbances can be minimized if a diffusing screen is placed in front of, or behind the phase object. In this case the unfocused twin image appears only as a diffuse background

Fig. 4.21 Wrapped phase of a liquid system

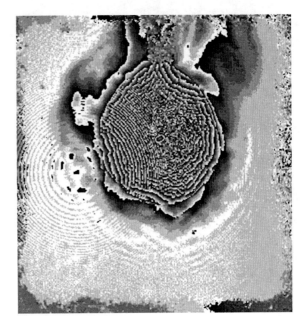

in the images, which does not disturb the evaluation. If a diffuser is introduced no additional tilting of the reference wave is necessary for image separation. A disadvantage of using a diffuser is the generation of speckle due to the rough surface.

In Fig. 4.21 a typical phase difference image (modulo 2π) of a transparent phase object is shown. The holograms are recorded with the set-up of Fig. 4.20 (without diffuser). The object volume consists of a droplet of toluene, which is introduced into the liquid phase water/acetone. The refractive index changes are caused by a concentration gradient, which is induced by the mass transfer of acetone into the droplet.

Chapter 5
Digital Holographic Particle Sizing and Microscopy

5.1 Introduction

While much of this book concentrates on the use of digital holography for vibration and stress analysis, contouring and metrology, the emphasis in this chapter is on the unique image forming characteristics of the hologram. Holography is inherently a high-resolution imaging system, and when coupled with its non-intrusive and non-destructive character and its preservation of parallax and perspective information of the original scene, it is ideally suited for observation, identification and mensuration of microscopic-sized particles, organisms and cell biology.

Following Gabor's original idea of a "wave-reconstruction microscope" in 1947 [68], Maiman's demonstration of the first working laser in 1960 [151] and its incorporation into holography in 1963 [142], holographic microscopy became a reality (but not quite in the form Gabor intended). It was recognised that holography was a powerful method for recording high-resolution images of microscopic organisms and particles, in three dimensions and in their natural environment. Because individual planes of the image can be spatially isolated (optically sectioned) by selective focusing of the reconstructed image, precise analysis and measurement of size, shape, distribution and concentration of microscopic particles is obtainable.

Applications quickly followed and holography was applied to sizing of air-borne particles [239], holographic microscopy in biomedical imaging [43], identification and distribution of marine organisms and floc in the water column [85, 117], and deployment from aircraft [242], to name but a few. Holographic cameras ("holo-cameras") were soon designed for deployment in laboratories, industrial plants and in inhospitable environments such as in space or subsea [24, 104, 118, 248]. These first-generation holocameras utilised, of course, "classical" analogue holography with recording on photosensitive silver halide emulsions coated on glass or film. They were bulky, heavy and difficult to manoeuver, however, with wet-chemical

© Springer-Verlag Berlin Heidelberg 2015
U. Schnars et al., *Digital Holography and Wavefront Sensing*,
DOI 10.1007/978-3-662-44693-5_5

processing of the photographic emulsions followed by precision reconstruction in dedicated laboratory facilities.

With the major advances in electronic image sensors towards the end of the 20th century coupled with parallel improvements in computer processing power and storage, digital holographic (DH) recording with numerical reconstruction by computer became a realistic possibility [198, 261]. Many of the advantages of classical holography still applied to DH, but now with the addition of rapid capture and storage, freedom from wet chemical processing, and, importantly, the ability to capture holographic videos ("holovideos") of moving objects which allowed the preservation of temporal as well as spatial dimensions. It is this ability to record true 3D, full-field, high-resolution, distortion free in situ images from which particle dimensions, distribution and dynamics can be extracted that sets digital holography apart from the other imaging methods and make it so useful for particle sizing and microscopy.

The concepts and techniques of particle sizing and imaging using classical holographic recording on photographic light-sensitive emulsions are well-documented in the literature, see e.g. [79, 240, 244]. Many of these concepts are equally applicable and relevant to digital holographic recording on electronic sensor arrays. We saw in previous chapters that in DH, the interference field is stored directly in computer memory and digitised in accordance with the pixel spacing of the electronic sensor. The holographic image is reconstructed by numerical simulation of the propagation of the optical field through space; planar sections of the image can be individually reconstructed on the computer monitor at any distance from the hologram plane, and in any time frame. This is analogous to refocusing the image plane in conventional microscopy, thereby allowing the size, shape, relative location, identification and distribution of particles to be extracted. Since reconstruction does not require a dedicated optical reconstruction facility, algorithms can readily include specialist techniques such as dark-field, phase reconstruction, and pre- and post-processing. Importantly, in DH the phase of the wavefield as well as its intensity is retained on reconstruction.

5.2 Recording and Replay Conditions

For digital holographic particle sizing, the 'in-line' (or 'on-axis') Fraunhofer mode (ILH) mode of recording is generally the favoured method of recording by virtue of its geometric simplicity, and consequently its lower cost and potentially high resolution. Because of the much higher sensitivity of electronic sensors compared with holographic photosensitive film, low-power continuous-wave (c.w.) lasers can be used for illumination. In many field applications though, pulsed lasers are often a better choice, particularly if the subject is fast-moving or subject to vibration, or the holocamera cannot be held steady with respect to the target. The depth-of-field in ILH is not strongly constrained by the laser coherence (since object and reference beams travel similar paths) and the main limitation is the reduction in resolution with distance from the sensor (this will be discussed in more detail later in this chapter). No lenses need be involved in image formation but can be used with advantage in some variations of the geometry to increase the magnification of the reconstructed images; the stability and accuracy of the sensor geometry provides in-built calibration.

5.2.1 In-line Recording

Digital holograms of particle distributions within a semi-transparent sample volume can be recorded with the in-line set-up depicted in Fig. 5.1. The basic digital ILH geometry employs a single collimated laser beam (a plane wave) directed through the semi-transparent sample volume towards the electronic sensor. The sensor is usually a 2D array such as a CCD (charge coupled device) or complementary metal oxide semiconductor (CMOS) device. Optical interference at the sensor plane occurs between light diffracted by the particles and the undiffracted (straight through) portion of the illuminating beam. An overall scene transparency of around 60–80 % is needed so that enough light reaches the sensor to record holograms with a good signal-to-noise ratio.

In particle sizing applications, the objects of interest may range in size from a few micrometres to a few millimetres, and we have to distinguish between recording the object in the "far-field" (a Fraunhofer hologram) or in the "near-field" (a Fresnel hologram) with respect to the sensor plane. It is normally assumed that the crossover between these two regimes occurs at a distance given by the "Fraunhofer far-field condition". A spherical particle of diameter 2a is deemed to be in the far-field if it is located in a plane, z_0, of object-space such that the distance between it and the sensor is given by

$$z_0 \geq z_F = \frac{(2a)^2}{\lambda} \tag{5.1}$$

Fig. 5.1 Recording geometry for in-line holography of small particles

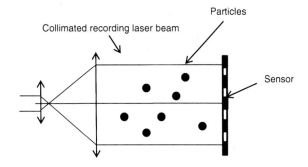

where z_F is the Fraunhofer far-field distance and λ is the illuminating wavelength. In practice, it is assumed that if the particle lies between about 1 and 100 far-fields from the sensor then good recording of the diffraction pattern (the well-known Airy disc) will be obtained (see Thompson [239]). This will result in a diffraction pattern of the particle at the sensor which will be unaltered in shape with distance but will increase in size the further it is away from the sensor. This distance we can call the "Fraunhofer range". For a 100 μm particle recorded at a wavelength of 532 nm, the sensor should lie between approximately 19 mm and 1.9 m from the object; but for a 1 μm target, the Fraunhofer range reduces to between 1.9 and 190 μm. Beyond 100 far-fields fringe contrast decreases to a level such that good recording is unattainable. For an object of 1 mm size, the minimum far-field distance stretches to nearly 2 m; if we do image in these conditions we have created a Fresnel diffraction pattern (near-field). Different Fresnel patterns are recorded at different distances from the sensor; however, the use of a suitable reconstruction algorithm will still allow focusing of particles. In laboratory implementations of classical holography, relay lenses are often employed to reduce the space implications and to increase image magnification (see Thompson [239]).

Images of any plane in the recording volume at specific distances from the sensor can be recreated and high-resolution dimensional measurement extracted from them. In ILH, only outlines of the original object are seen (if it is non-transparent) and because the sensor area is small the images have almost no parallax. If the beam diameter is larger than the sensor, objects located outside the direct sampling volume can contribute to the interference. The replay geometry of a classical ILH and the location of the images obtained are shown in Fig. 5.2. Two images are reconstructed: one between the sensor and the laser (the primary or virtual image) and one in front of the sensor between it and the observer (the secondary or real image), and at the original recording distance. Of course in digital replay we only reconstruct in one plane at a time: an image of a particular plane is viewed on a computer monitor and represents the virtual or the real image depending on the image distance which is fed to the computer algorithm. An out-of-focus conjugate image background is always present whichever image is reconstructed: the two images are optically indistinguishable from each other. From

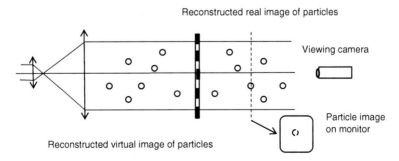

Fig. 5.2 Replay geometry of a classical in-line hologram

the diagram we can see that the real image is "pseudoscopic" i.e. the image is reversed back-to-front and right-to-left but of course this makes no difference to the computer reconstruction.

5.2.2 Off-axis Recording

To record large, opaque or dense aggregates of particles, the off-axis holographic mode (OAH), also known as side-reference beam or twin beam, is more suitable. Because the recorded subjects are generally larger than in in-line, the far-field condition is not usually satisfied and these are classified as Fresnel holograms. The geometry is more complex since it involves front or side illumination of the subject with a separate off-axis (angular) reference beam. The laser beam is amplitude divided using a beam-splitter into two beams that travel different paths to the sensor (see Figs. 2.10 and 4.20). One beam, the reference, is directed onto the sensor without impinging on the object; the second beam, the object beam, is used to illuminate the scene to be recorded (the object beam may be further subdivided to illuminate from a variety of directions) and it is the scattered or reflected light which arrives at the hologram plane and interferes with the reference beam. A major advantage of the off-axis mode is the angular separation of the reference and object beams. On classical replay we can select virtual or real image reconstruction of the scene by choice of reference beam parameters and angular orientation. In classical holography the reference beam is usually incident on the hologram film at an oblique angle of as much as 60°. However, in DH because of the smaller pixel dimensions of electronic sensors (up to 2 or 3 μm for CMOS compared with grain sizes of 30 nm or so with holographic film, see Table 3.1) the reference beam angle is restricted to 10° or less (or even at virtually zero angles).

Off-axis recording configurations are commonly used in Holographic Particle Image Velocimetry (HPIV) and Holographic Microscopy, outlined later in this chapter.

5.2.3 Image Resolution

The computer algorithms that we saw earlier being applied to digital holography (see Chap. 3) are equally applicable to, and suitable for high-resolution imaging applications like particle sizing. Both the Fresnel approximation and the convolution algorithms (several variations such as the "angular spectrum" method bear a close resemblance to the convolution approach and produce similar results) render the Fresnel-Kirchhoff equation more amenable for digitisation and computer implementation [28, 130]. No matter which approach is used, the intensity and phase distribution of the wavefield at any plane in the reconstructed volume of the hologram is recreated, thereby simulating the effect of traversing an image sensor or refocusing of the image plane through an optically replayed hologram, but with the added advantage that phase information can also be extracted, if needed. The main differences between the methods are in the speed of processing, scaling of the reconstructed images and their applicability to ILH or OAH recording.

An important feature of the Fresnel approximation (Sect. 3.2) is that the extent of the reconstructed image (effective $\xi' - \eta'$ dimensions in the image plane) increases with distance from the sensor. Accordingly, the sampling interval at the reconstructed plane also needs to increase with distance from the hologram thereby reducing the effective image resolution in the image plane. The direct in-line portion of the reconstructed image occupies fewer pixels as the reconstruction distance increases, and is surrounded on all sides by any components which are outside the zero-order beam path (straight through). The consequence of this is that as reconstruction extends further from the hologram plane, the effective pixel pitch increases, covering a larger area in image-space. For N square pixels of dimension Δx in the x- and y-directions the required sampling interval is given by (see Eq. 3.23)

$$\Delta \xi = \frac{\lambda z_i}{N \Delta x} \tag{5.2}$$

where z_i is the image-sensor distance and $N \Delta x$ is the approximate overall dimension of the sensor.

A pyramidal volume is reconstructed extending out from the hologram, but the resolution approaches that given by Eq. (5.2). This behaviour is similar to that of a classical imaging system employing a lens, except that in the holographic case, the viewing angle is dependent on the maximum spatial frequency of the sensor and hence the maximum beam angle that can be recorded by the sensor (off-axis). Diffraction-limited resolution of a holographic real image, in the absence of all aberrations, is usually defined as the ability to distinguish two points in the image plane separated by a distance r in the transverse plane normal to the optic axis and is given by, for example, Born and Wolf [16],

$$r = 1.22 z_i \frac{\lambda}{D} \tag{5.3}$$

where D is the diameter of the aperture. Save for the constant factor (1.22), the similarity between Eqs. (5.2) and (5.3) is apparent. We can interpret $\Delta \xi$ as being equivalent to the diffraction limited resolution of the system in the transverse plane (equivalent to r). For square pixels of 3 µm dimension, N of 1,024, a wavelength of 532 nm and z_i of 100 mm, $\Delta \xi$ is about 17.5 µm. Note also, though, that as we approach a resolution equal to that of the pixel spacing, the (Nyquist) sampling theorem will come into play. From it we can deduce that the best achievable resolution can be no better than twice the pixel dimension, i.e. 6 µm in this case. Given these characteristics, the Fresnel approximation lends itself more to the reconstruction of off-axis holograms or in-line holograms with a significant number of particles/points which are outside the extent of the zero-order area (see also the discussion in Sect. 3.2.2).

For in-line holography of microscopic particles it is preferable that the pixel scale be maintained over long path lengths so as to allow for easy comparison of images between planes by reconstructing only the in-line portion of the hologram at each plane of interest. Using the angular spectrum [130] or convolution approach (Sect. 3.2.2), the number of sample points, and the spatial sampling interval (pixel pitch) is unchanged regardless of reconstruction distance z_i; however, the algorithm will only reconstruct components within the confines of the zero-order term. Once again though we must take the sampling theorem into account so again our best resolution is $2\Delta x$. For reconstruction of collimated beam in-line holograms, the spatial sampling interval used in the algorithm is equivalent to the detector pixel spacing and on reconstruction the same pixel spacing is maintained, such that according to Eq. (3.35)

$$\Delta \xi = \Delta x \qquad (5.4)$$

This would appear to indicate that we could achieve a higher resolution using this approach, however, the Fresnel approximation reconstructs to the physical diffraction limit of the system and therefore the convolution approach provides only an apparent increase in resolution. It is still the diffraction-limited resolution (Eq. 5.2) that defines system performance.

There are another few advantages to the convolution approach: firstly unwanted frequencies may be eliminated after transformation to its spectrum and before the image is reconstructed [48]. A second advantage is that because there is no practical minimum reconstruction distance imposed before the algorithm breaks down, it therefore allows the entire hologram volume to be scanned [152].

The sensor, whether it is photographic or electronic, must be able to record the interference pattern resulting from the hologram: thus the pixel dimensions and the overall extent of the sensor have a great bearing on imaging capabilities. We saw from Eq. (3.53), that the maximum recording beam angle (in radians) for an off-axis hologram is given as

$$\theta_{max} \approx \frac{\lambda}{2\Delta x} \qquad (5.5)$$

Even with the best (current) available sensor of 1.4 μm pixels, the maximum recording angle for digital holography is restricted to about 11° with a consequent restriction in the field-of-view, compared with angles of 45° and more with classical off-axis recording on photofilm (see also Table 3.1). To record a moving particle a rule of thumb often applied is that the object must not move by more than one-tenth of its diameter during the exposure. Thus for a 10 μm particle moving at 1 ms^{-1}, we would need a maximum exposure duration of about 1 μs.

For off-axis holograms the presence of speckle introduced by the coherence of the light and the finite aperture of the viewing system sets a lower limit to the achievable resolving power of the system. In practice, the minimum resolving power is increased by a factor of two to three to take this into account.

5.2.4 Holographic Depth-of-Field and Depth-of-Focus

The terms depth-of-field and depth-of-focus are well-known concepts in imaging optics, but can be misunderstood and often used interchangeably. In conventional imaging with a lens, field-depth is usually taken to mean the axial range in object-space over which an image of a scene provides acceptable resolution in a given image-plane. The depth-of-focus, by contrast, is the corresponding axial range in image-space over which the object scene is deemed to be in acceptable focus. For a camera lens focused at its hyperfocal distance, the depth-of-field essentially stretches from the closest focusing distance to infinity. In lens-based microscopy the recordable depth-of-field is restricted to about λ/NA^2 for a spherical particle (NA is the numerical aperture of the lens) [81]; for a microscope objective of 0.1 NA, the depth-of-field is about 50 μm at a wavelength of 500 nm and the focus depth is of the order of 100 μm or so. When we contrast this with holography the situation changes dramatically, and we have to redefine the concepts slightly.

One useful measure in holography we can apply is the "Fraunhofer range" discussed earlier (Sect. 5.2.1). This we can interpret as the depth-of-scene in object-space over which any particles can be recorded and still fulfil the Fraunhofer condition. From our previous argument we can see that this is effectively the range between 1 and 100 far-field distances of the object.

For 100 μm particles recorded at 532 nm, the Fraunhofer range is about 1.9 m. This distance would normally exceed any laser to sensor path length in a typical holocamera.

A more realistic measure of depth-of-field is obtained by defining it in terms of the axial range in object-space where a particle of a given diameter could be located and still be reconstructed at its conjugate image plane with acceptable resolution. Here we assume recording and replay at the same wavelength with a collimated reference beam, and that $z_0 = z_i$. In this situation a generally applied rule-of-thumb is that for "good" recording of the diffraction pattern of a spherical particle so that it can be reconstructed at an acceptable image resolution, the sensor must be large enough to capture the central maximum of the Airy profile plus 3 side-lobes. Following the treatment of Thompson [239] or Hariharan [79], and assuming a circular occlusion (i.e. particle) of diameter $2a$ and including a factor of 4 to account for the side-lobes, we can write Eq. (5.3) in terms of the radius out to the third side-lobe as

$$r_{\max} \approx 4z_0 \frac{\lambda}{2a} \qquad (5.6)$$

Interpreting r_{max} as the radius of the sensor needed to capture three side-lobes of the diffraction pattern of the particle, we can see that the distance, Δz_0, in object-space over which a good image in the above sense can be recorded is

$$\Delta z_o \approx \frac{(2a)r_{max}}{4\lambda} = \delta_o \tag{5.7}$$

This we can interpret as the depth-of-field. For the previous 100 μm particle and a sensor radius of 6 mm, the depth-of-field is about 280 mm. Contrast this with the depth-of-field obtainable using conventional microscopy. This we can interpret as the axial range in object-space over which a particle of a given diameter can be located and still provide a good image in the corresponding image plane.

However, since in holographic microscopy and sizing, we reconstruct particles at specific planes throughout the image-space, the separation between individually selectable planes is important: this is governed by the focus depth over which the image provides an acceptable resolution. The depth-of-focus, therefore, for an in-line hologram can be defined as the axial distance in image-space over which an image of a small particle located in object-space at a plane, z_o is formed at a distance z_i, and possesses acceptable resolution; in other words it represents the maximum defocusing distance from the exact diffraction limited image of a point-source. As we have seen before for a particle of diameter $2a$, an image of radius r will be formed at a distance $z_0 = z_i$ from the sensor in accordance with the Airy disc formulation. As we move the reconstruction plane axially towards or away from the sensor, the peak intensity of the central maximum will fall. Using another rule-of-thumb we assume that for a point source in a diffraction-limited system 80 % of light energy stays within the Airy spot when the image is defocused by $\pm\delta_I/2$. Following the treatments of Vikram [244], Meng and Hussain [156] or, Born and Wolf [16] we can write the total focusing distance around the image plane as,

$$\delta_I \approx \lambda \left(\frac{z}{r}\right)^2 \tag{5.8}$$

and substitute for r into the Airy disc formulation [Eq. (5.3)] to arrive at

$$\delta_I \approx \frac{(2a)^2}{\lambda} \quad \text{Depth-of-Focus} \tag{5.9}$$

as an approximate expression for the image depth over which resolution can be maintained (which is of the same order of magnitude as the far-field distance). For a 100 μm particle, the amount of defocus is about 19 mm. We can interpret this as the distance in image-space over which, on reconstruction, the diffraction pattern begins to be recognisable as a particle, through its best focus and on until it is just recognisable again.

We should note though that the above relationships are based on the assumption of recording small spherical particles in diffraction-limited systems. Focus depth is the range over which spherical particles appear to be same size as the original. In many DH systems the targets are much more complex and irregular, such as subsea plankton (see later in chapter) and in such systems focus depth relates more to the smallest desired resolvable feature e.g. hairs on a copepod.

5.2.5 Optical Aberrations

The usefulness of holography for accurate inspection, dimensional measurement and particle sizing is dependent on its ability to reproduce an image of a subject which is both low in optical aberrations and high in image resolution over a large depth-of-field. Resolution, contrast and noise are the dominant factors in a holographic image for particle measurement and identification, rather than brightness. In practice, image degradation can occur at any stage in the recording and replay of a hologram. All the primary monochromatic aberrations found in any optical system (spherical aberration, astigmatism, coma, distortion and field curvature) may be seen in a holographic image (see Sect. 2.6.2). Meier [155] developed a series of relationships which allowed the magnitude of the primary aberration coefficients to be calculated for holographic recording of point sources; these were later extended to non-paraxial imaging by Champagne [27], Latta [139] and others. Many optical ray tracing simulation programs now allow for the evaluation of aberrations in holography.

From the Meier relationships (see Eqs. 2.64–2.73) we see that the coefficients of spherical aberration, coma, astigmatism, field curvature and distortion can all be defined in terms of the location of the object and reference sources, the beam curvatures and the recording and replay wavelengths. Analysis of these relationships shows that aberration-free optical reconstruction (of the virtual image) can only be obtained if the hologram is replayed by an exact replica of the original reference wave, in terms of its curvature, wavelength and beam coordinates. For optical reconstruction of the real image, replay should be accomplished using an exact phase conjugate of the recording reference wave (same wavelength, opposite divergence and opposite direction); then the lateral, longitudinal and angular magnifications of the real image will all equal unity and aberrations will be reduced to a minimum (see Sect 2.6.2). In any case, because in ILH the object and references beams travel roughly collinear paths the dominant aberration is spherical and the others can be generally ignored.

In digital holography, aberration compensation can be added to the reconstruction algorithms and its application to several different classes of digital hologram recording geometry can be seen in Claus et al. [30].

5.3 Data Processing and Autofocusing of Holographic Images

A recurring issue in all forms of holographic recording (digital or classical) of microscopic particles, and one which imposes limits on the wider use of the technique, is extracting, processing and analysing the vast amount of data contained in the hologram. Although analysis can be performed by manual scanning of the reconstructions, this is tedious and time-consuming, and requires high levels of concentration; automatic interrogation of the data is essential for all practical applications of the technique.

Since a single holovideo may contain as much as several gigabytes of data, a first step in interpretation of digital holograms and holovideos of particles often involves localising every particle within a frame (in x,y,z,t), and distilling particle shape and positional information from it. This facilitates application-specific processing, with subsequent image recognition, particle tracking, counting and sizing. For each video frame, the hologram is reconstructed in incremental steps and an image of each slice parallel to the sensor plane is sequentially obtained. This is equivalent to discretised simulation of the wave-field projected into real image space by an analogue hologram when reconstructed by an optical beam. When a particle coincides with a reconstruction plane it will appear in-focus, and is characterised by a maximisation of image gradients at the particle edges.

Several methods of focus detection in DH have been implemented and reported in the literature; a non-exhaustive list includes self-entropy [115], amplitude analysis [49], Fresnelet-sparsity [144], "l-1" norms [145], and gradient measurement [21, 186, 194]. Focus detection is typically the first step in hologram analysis and with suitable software, particle identification can be carried out on focused particle images and 3D plots of relative position and distribution produced [153, 178, 221]. Most of these algorithms implement optimal-focus metrics which depend variously on image intensity gradient, variance, correlation, histograms and frequency-domain analysis. These metrics all rely on the premise that focused images have higher information content than blurred (out-of-focus) images due to the existence of larger gradients with higher variance across them. This leads to a greater deviation between maxima and minima in the brightness histogram and the local maximisation of power in higher frequency components when the image is transformed to the frequency-domain. Due to speckle in the hologram an algorithm with good noise immunity is required. One such algorithm is the contour gradient algorithm [22, 249] which has been used to analyse a number of holographic videos recorded in the North Sea (see Sect. 5.4.6). Figure 5.3 shows a hologram frame recorded on a 10.5 μm pixel pitch, and the associated contour outputs generated for the frame; three particles found in a single frame are shown. This algorithm achieves maximum noise immunity by constraining gradient measurements to particle edges.

5.4 Some Applications in Imaging and Particle Sizing

Having established the fundamental properties of high resolution imaging for particle sizing we can now outline some of the more common applications of the technology. Firstly, though, a note on terminology: in all of the applications outlined here, we are dealing with microscopic particles ranging in size from a few micrometres to a few millimetres and all such uses of digital holography for particle sizing at micrometre scale can be considered as digital holographic microscopy (DHM). The term, however is commonly used to describe its use in applications

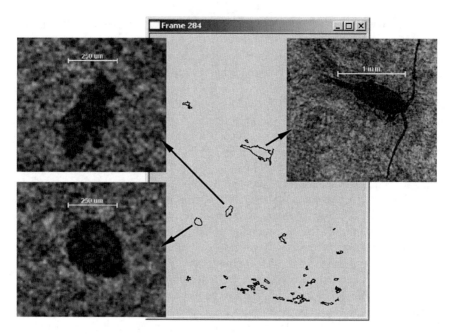

Fig. 5.3 Images of reconstructed particles using contours to locate them in a holovideo

synonymous with conventional microscopy, such as for biomedical imaging of cell tissues. In this section we will restrict the term digital holographic microscopy (DHM) to this type of system. Furthermore many applications are now concerned with movement of particles and measurement of velocity vectors and this type of system could also be considered as a variation of Holographic Particle Image Velocimetry (HPIV).

5.4.1 Particle Sizing

The pioneering work in particle sizing was accomplished by Thompson (see [239]) and co-workers in the 1960s. The work of Thompson and others utilised the concepts of in-line Fraunhofer recording, the basics of which were previously outlined earlier in this chapter. The general geometry of Fig. 5.1 was used to record holograms of air-borne particles. Later an imaging device known as a "disdrometer" was developed for the measurement of the size of fog droplets down to about 4 μm dimension. In this system use was made of a telescope imaging system which relayed a magnified image of a plane to the holographic film thereby providing an increase in effective resolution.

Many workers have since applied digital Fraunhofer ILH to sizing of atmospheric particles. Figure 5.4 shows a digital hologram of atmospheric particles and

Fig. 5.4 A hologram of in-situ atmospheric particles (*upper left*); the other images are reconstructions of the hologram at distances of 105.5, 123.2 and 141.7 mm respectively from the sensor plane (courtesy of Raupach [191])

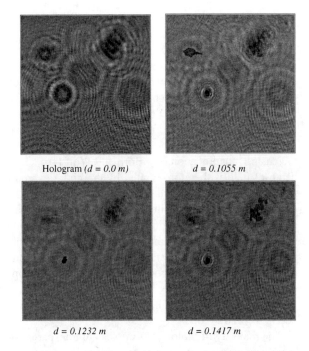

Hologram *(d = 0.0 m)*   *d = 0.1055 m*

d = 0.1232 m   *d = 0.1417 m*

its reconstruction in a PC at reconstruction distances of 105.5, 123.2 and 141.7 mm from the sensor plane [191].

5.4.2 Digital Holographic Microscopy (DHM)

In DHM, illumination is commonly accomplished by replacing the collimated reference beam by a divergent beam (Fig. 5.5). Utilising this geometry can increase the magnification of the image, with an apparent improvement in resolution; but this is at the expense of a reduced sampling volume. In Fig. 5.5 the incident beam is focused through a pinhole and diverges onto the sensor a distance R away from the pinhole.

Fig. 5.5 Divergent beam recording

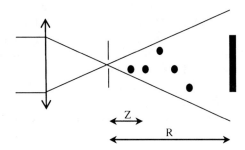

Reconstruction is required only over a small fraction $1/M^2$ of the hologram area, where M is the magnification factor Z/R, and Z is the reconstruction distance [156]. As the reconstruction plane is moved closer to the centre of divergence of the recording beam the reconstruction area decreases with an apparent improvement in spatial resolution; this apparent increase in resolution comes at the expense of recording volume. This scheme is used in several of the subsea holocameras that have been developed. Digital Holographic Microscopes often utilise off-axis recording to enable their use in either transmission or reflection [43].

As we saw earlier, the depth-of-field of an imaging system decreases with increasing magnification (see Born and Wolf [16]). In conventional microscopy the depth-of-field is therefore limited by the high degree of magnification that is often needed. Investigation of a three dimensional object with microscopic resolution requires continual refocusing of the image plane to maintain the resolution over the depth of the object. Digital Holography facilitates focusing at different object planes. In addition, the images are free of optical aberrations brought about by imperfections of optical lenses in the image formation path. Fundamental work in Digital Holographic Microscopy (DHM) and its application to measurement of micro-mechanical systems has been carried out by several workers [31, 78, 107] and DHM is further finding increasing application in biology and medicine [43, 108, 114, 226].

The majority of biological samples are transparent or semi-transparent. For such specimens phase contrast is of particular importance. Conventionally phase contrast images are generated only with the special technique of phase contrast microscopy. With DHM phase images are available directly as a result of the numerical reconstruction process.

In order to obtain a high lateral resolution in the reconstructed image the object should be placed near to the electronic image sensor (see Chap. 3). The necessary distance to obtain a given resolution $\Delta\xi'$ with the Fresnel approximation can be estimated using Eq. (3.23 or 5.2), i.e.

$$\Delta\xi' = \frac{\lambda d'}{N\Delta x} \tag{5.10}$$

As before, the prime denotes parameters in the reconstruction plane (in holographic microscopy, the reconstruction distance d' may be different from the recording distance d). For a pixel size of 10 μm, a wavelength of 500 nm, 1,000 pixels in the x-plane and a desired resolution of $\Delta\xi' = 1$ μm, a reconstruction distance d' of 2 cm is necessary. At such short distances the Fresnel approximation is no longer valid and the convolution approach to reconstruction is more appropriate. However, the resolution of an image derived from the convolution method approximates to that of the pixel dimensions of the sensor (see Eq. 5.2); in this case about 10 μm, which is too low for microscopy. Therefore the reconstruction procedure has to be modified.

The lateral magnification of the holographic reconstruction can be derived from the holographic imaging equations (see Sect. 2.6.2). According to Eq. (2.70) the lateral magnification of the reconstructed virtual image is:

$$M = \left[1 + \frac{d}{d'_r}\frac{\lambda_1}{\lambda_2} - \frac{d}{d_r}\right]^{-1} \qquad (5.11)$$

where d_r and d_r' describe the distances between the point source of a spherical reference wave and the hologram plane in the recording and reconstruction process, the recording and reconstruction wavelengths are λ_1 and λ_2 respectively. The reconstruction distance d' i.e. the position of the reconstructed image, can be re-written as,

$$d' = \left[\frac{1}{d'_r} + \frac{\lambda_2}{\lambda_1}\frac{1}{d} - \frac{1}{d_r}\frac{\lambda_2}{\lambda_1}\right]^{-1} \qquad (5.12)$$

If the same reference wavefront is used for recording and reconstruction it follows that $d' = d$ (note that d, d', d_r and d_r' are always counted positive in this book).

Magnification can be introduced by changing the wavelength or the position of the point source of the reference wave in the reconstruction process. If the desired magnification factor is determined, the reconstruction distance can be calculated by a combination of Eqs. (5.11) and (5.12) with λ_1 set equal to λ_2, so that,

$$d' = d \cdot M \qquad (5.13)$$

To enlarge the image, the source of the reference wave needs to be placed at a distance

$$d'_r = \left[\frac{1}{d'} - \frac{1}{d} + \frac{1}{d_r}\right]^{-1} \qquad (5.14)$$

The reference wave can now be described by,

$$E_R(x, y) = \exp\left(-i\frac{2\pi}{\lambda}\sqrt{d_r'^2 + (x - x_r')^2 + (y - y_r')^2}\right) \qquad (5.15)$$

where $(x_r', y_r', -d_r')$ is the position of the reference source point in the reconstruction process.

A simple set-up for digital holographic microscopy is shown in Fig. 5.6. The object is illuminated in transmission and the spherical reference wave is coupled into the set-up via a semi-transparent mirror. Reference and object waves are guided via optical fibres. For weak scattering objects the external reference wave can be blocked and a conventional in-line configuration used. A digital hologram of a test target recorded using this set-up is shown in Fig. 5.7a. The corresponding intensity reconstruction is depicted in Fig. 5.7b. The resolution obtained corresponds to about Group 4, element 3 (about 2.2 µm).

Depeursinge [43] shows two more complex DHM configurations currently being used for biomedical cell studies; one based on transmission microscopy and the

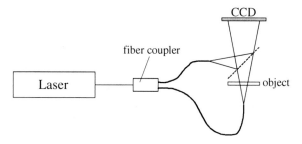

Fig. 5.6 Basic digital holographic microscope layout

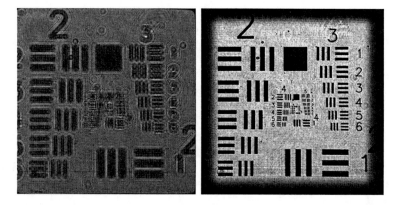

Fig. 5.7 Hologram of resolution target and its numerical reconstruction (recorded in system similar to Fig. 5.6)

other on reflection techniques. Figure 5.8a shows a transmission scheme and Fig. 5.8b shows a DHM in reflection mode.

5.4.3 Holographic Tomography

As indicated earlier, the focused image depth of an in-line hologram can be of the order of tens of millimetres and makes the localisation of particles difficult in the replayed image. Particle localization along the optical axis is only possible within a defined range given by Eq. (5.10). Adams et al. [5] and Kreis et al. [127] showed that improved particle localization could be obtained by, for example, combining DH with tomographic methods. The following description is based on that of Adams et al. [5]. In tomography several projections in different directions through a scene are recorded. The three-dimensional distribution of the physical quantity, e.g. the attenuation of a beam passing the scene, is then calculated by numerical methods. To record simultaneous multiple in-line holograms of the particle stream

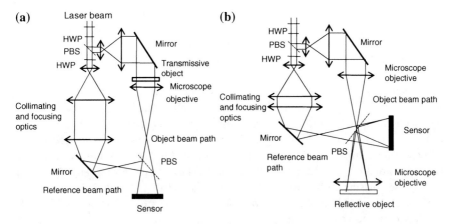

Fig. 5.8 Transmission (**a**) and reflection (**b**) digital holographic microscope configurations (adapted from [43]). PBS and HWP represent polarizing beam splitters and half-wave plates, respectively

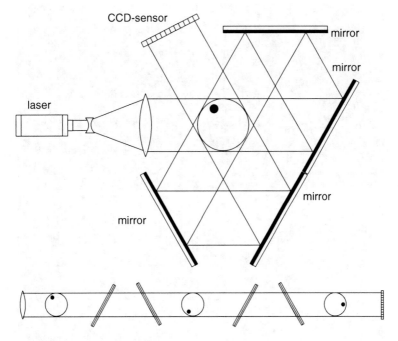

Fig. 5.9 Experimental set-up for recording in-line holograms from several directions with one CCD-sensor and deconvoluted lightpath (from [5])

from different directions an arrangement using a single sensor array is used, (see Fig. 5.9). After passing the particles a first time, the collimated beam of a ruby laser is guided by two mirrors through a second pass and by two further mirrors through a third pass along the stream before reaching the sensor. The lower part of Fig. 5.9 shows schematically the effective location of the mirrors and the deconvoluted light path. Now it is possible to extract the three views of the particles by three reconstructions with numerical focusing to the different planes at 40, 65.5 and 95.5 cm from the sensor.

Particles with a size of 250 μm were distributed randomly throughout the volume. The CCD sensor comprised an array of 2,048 × 2,048 light-sensitive pixels with a pitch of 9 μm × 9 μm. The diffraction rings of each particle can be seen in the in-line holograms (Fig. 5.10). In addition to the diffraction rings caused by the particles, the image overlaps with a pattern produced by the recording system itself. On reconstruction, the particles are visible as dark spots without any diffraction rings. The missing diffraction rings or halos show that the particles are reconstructed at the correct distance. The measured average particle diameter is 28 pixels, corresponding to about 250 μm.

The three reconstructed images show the particle stream from different directions. To gain a three-dimensional particle distribution from these images, a

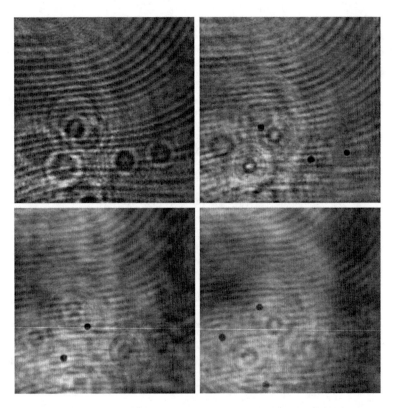

Fig. 5.10 In-line hologram, recorded with the set-up of Fig. 5.9 (*upper left*). The other images show reconstructed particle distributions at a distances of 40, 65.5 and 95.5 cm, respectively (from [5])

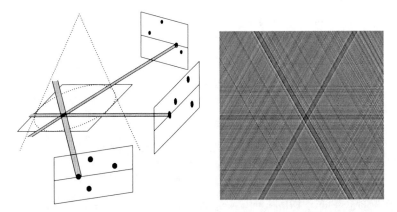

Fig. 5.11 Principle of tomography. From the existing images single lines are taken and combined to a two-dimensional distribution by methods of tomography. *Right* two-dimensional distribution gained from three different views. In the crossing of the stripes a particle is found (from [5])

tomographic method is applied, (Fig. 5.11). From every image in Fig. 5.9 a line is drawn back to the hologram, and from these three lines a two-dimensional profile through the particle stream is calculated with a method based on the filtered back projection approach of tomography. If this method is applied for all lines in the reconstructed images, a full three-dimensional distribution of the particles is achieved. The three back projections of one particle must be in one plane. In the crossing of the three stripes a particle is reconstructed. The stripes arise from the low number of views used in the tomographic evaluation. With increasing number of views the stripes begin to disappear.

5.4.4 Phase Shifting DHM

Digital Holography can also be combined with conventional microscopy using high aperture lenses as imaging devices. Such a set-up for investigating samples in transmission is shown in Fig. 5.12. A light beam is coupled into a Mach-Zehnder interferometer. The sample to be investigated (the object) is mounted in one arm of the interferometer. It is imaged onto the CCD/CMOS sensor by a microscope objective (MO). A second objective is mounted in the reference arm in order to form a reference wavefront with the same curvature. Both partial waves interfere at the CCD target. An image of the sample superimposed by a coherent background (reference wave) is formed onto the CCD target.

A set of phase-shifted images is recorded [255, 256]. The phase shift is realized, for example, using a piezoelectric transducer (PZT) in the reference arm of the interferometer. From these phase shifted images the complex amplitude of the object wavefront in the image plane can be calculated as described in Sect. 3.3.3.

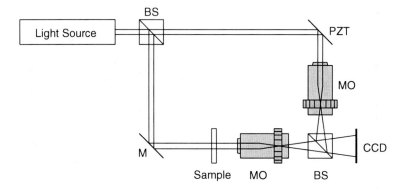

Fig. 5.12 Phase-shifting DHM with image plane recording

Numerical refocusing into any other object plane is now possible using the Fresnel-Kirchhoff integral.

A few slightly different methods are described in the literature. An off-axis recording geometry can be realized, if both beams are slightly tilted with respect to each other. In this case phase shifting is not necessary to calculate the initial phase. The algorithms described in Sect. 3.2 can be used directly to calculate the complex amplitude in other planes. Similar set-ups to that depicted in Fig. 5.12 have been used for recording of holograms in reflection.

The quality of images recorded with coherent light is in general worse than those recorded with incoherent light due to coherent noise. Dubois et al. [47] developed therefore a phase shifting digital holographic microscope with an ordinary LED as light source. The image quality improves (less speckle noise) due to the reduced spatial coherence of that light source compared to images generated by a laser.

5.4.5 Particle Image Velocimetry (PIV)

Another important area where DH is beginning to play a major role is in fluid dynamics where currently particle image velocimetry (PIV) is the favoured method of flow visualization to obtain instantaneous velocity fields and measurements. In traditional PIV, the fluid is "seeded" with tracer particles which, for sufficiently small particles, are assumed to flow with the general streamlines of the fluid. The seeded fluid is illuminated with a sheet of laser light to make the particles visible to the detector system. The lightsheet is viewed normal to the fluid flow with a digital camera and dedicated software applied to visualize the vector maps.

The key element that digital holographic PIV (D-HPIV) brings is the ability to capture the velocity field in all three spatial dimensions and also with the added bonus of the time dimension. The intensity field is interrogated using 3D cross-correlation techniques to yield the velocity field. Importantly in-line holography utilises forward scattered light, rather than side scattered light as used in conventional PIV, to increase the sensitivity of the method by a few orders of magnitude [10, 83, 213]. An excellent overview of HPIV is given in a Special Issue of Measurement Science and Technology of 2003 [84].

Meng et al. [157] outline a variety of potential effects which can degrade the quality of digital HPIV information. However we should note that many of these effects are not confined solely to HPIV but apply to holographic particle sizing in general. Such factors include the assumption that the recorded particles are small and spherical, and the influence of speckle which reduces the signal-to-noise ratio of particle images. Furthermore noise can also be introduced through impurities in the scattering medium, such as temperature variations and window blemishes. The combination of these factors increases the complexity of the correlation process.

5.4.6 Underwater Digital Holography

When the concepts of DH are applied underwater, we have an imaging technique which affords marine scientists the opportunity to study the aquatic environment in a way never before possible. Knox's seminal paper in 1966 [117] acted as the catalyst for the growth of holography in marine science. It sparked the development of a series of submersible holocameras which were deployed in the oceans around the world. These early holocameras demonstrated the potential of holography for imaging and measurement of marine and freshwater organisms and particles down to a few micrometres dimension. However, these holocameras were bulky, heavy and difficult to manoeuvre from ships or operate from remotely operated vehicles (ROVs); and deployment was restricted to a few hundred metres below the surface. Furthermore, the holograms needed wet-chemical processing and the consequent reconstruction in a dedicated replay facility. While impressive images were obtained by these systems, the time-consuming and laborious data extraction procedures limited the amount of meaningful scientific results that were obtained. Furthermore, the gradual withdrawal of holographic materials from the market place led to the near demise of classical holography.

The dramatic improvement in electronic sensor arrays, availability of more compact lasers and the vast increase in computer power since the late 1990s led inevitably to the development of the first submersible digital holographic camera; by Owen and Zozulya [176] in 2000. It was based on the standard in-line Fraunhofer geometry shown in Fig. 5.1. It utilised a 10 mW continuous wave (c.w.) diode laser in an in-line configuration onto a CCD array over a maximum 25 cm path length. The use of a c.w. laser can be justified on the grounds that the enhanced sensitivity of electronic sensors over photofilm reduces exposure times to about 100 μs. However, this still limits application to slowly moving systems. This holocamera was successfully deployed in the field in Tampa Bay, Florida to depths of about 50 m.

Since this pioneering holocamera, several workers world-wide e.g. [97, 165, 214, 227, 228] have exploited DH for underwater environmental science, and submersible digital holocameras are now becoming commercially available [210]. Although these systems share common elements such as use of the in-line geometry and, usually, charge-coupled device (CCD) arrays, there are subtle differences amongst them relating to their application and deployment method. Many were developed primarily for phytoplankton studies over sampling volumes of around 1 mm^3–1 cm^3; and primarily low particle velocities. In most underwater applications, the objects (or the camera) are in motion during the exposure. The effect of this motion is to blur the finer fringes, and thus reduce resolution and contrast. In-plane motion is the most severe case, and adopting the experimentally-verified criterion that the maximum allowable motion is less than one-tenth of the minimum required fringe spacing of the smallest object, then for in-line holograms the maximum object motion must be less than one-tenth of the object's dimension. For particles of 10 μm dimension, and a typical Q-switched YAG laser pulse duration of 10 ns, a transverse velocity of up to 100 ms^{-1} can be tolerated. Off-axis holograms are more demanding in their requirements and the maximum allowed velocity is reduced by about an order of magnitude. This is, however, more than adequate for most field applications of the technique. When optical holograms are recorded in water and replayed in air, the refractive index mismatch between recording and replay spaces will cause the aberrations to increase. In ILH, only spherical aberration is significant since the object and reference beams travel very similar paths. However, in OAH astigmatism and coma dominate and both increase with the field angle of the subject in the reconstructed image. These limit resolution and introduce uncertainty in co-ordinate location. Furthermore, the water itself may be expected to influence the quality of the images produced. An increase in the overall turbidity of the water will adversely affect both in-line and off-axis techniques and would be expected to create a background noise that will reduce image fidelity.

One practical solution, unique to holography, compensates for the change of effective wavelength of light as it passes through water, by allowing a deliberate mismatch of recording and replay reference beams to off-set the refractive index mismatch [109, 110]. For holograms recorded and replayed in air, a usual prerequisite is that the reconstruction wavelength is the same as that used in recording. From the dependence of wavelength on the refractive index, n, we can apply the more general condition that it is the ratio λ/n that must remain constant. This relationship suggests that a hologram, immersed in water and recorded at a specific in-air wavelength will produce an aberration-free image in air (provided of course that the rest of the Meier conditions are complied with) when replayed at a reconstruction wavelength which is itself equivalent to the effective wavelength of light when passing through water. For example, if a green laser (532 nm) is used in recording in water, the ideal replay wavelength in air is around 400 nm (i.e. 532 nm divided by 1.33, the refractive index of water). However, complete correction assumes that the entire recording system is located in water; but since this is impractical, holograms are usually recorded with the holographic sensor behind a glass window. The additional glass and air paths affect the compensation of aberrations. However, third-order aberration theory (e.g. [81]) shows that if the window-to-air path length ratio is appropriately chosen for a specific replay wavelength, then aberration balancing will occur and residual aberrations are reduced to a minimum over a wide range of field angles and object locations. This behaviour can also be simulated using an optical ray-trace or optical design program. However, here again the advantages of DH come to the fore: the whole process of aberration compensation can be accomplished by incorporating it into the reconstruction algorithms.

The *eHoloCam* system was developed at the University of Aberdeen [114, 157] and differs from most other digital holocameras in using a pulsed frequency-doubled Nd-YAG laser to freeze any motion in fast moving objects. Holograms are recorded on a high-resolution CMOS sensor to record water volumes of around 36 cm^3 at video frame rates between 5 and 25 Hz. eHoloCam and its internal layout is shown in Fig. 5.13. It comprises of two water-tight housings. One housing (724 mm length by 330 mm diameter) contains the laser, on- board computer, two 320 GB SATA hard drives and beam forming optics. The laser is a pulsed frequency-doubled Nd-YAG laser operating at 532 nm, with 1 mJ per pulse over a 4 ns duration and pulse repetition rate up to 25 Hz. Another housing (170 mm length by 100 mm diameter) contains a 10.50 mm × 7.73 mm area CMOS sensor with 2,208 × 3,000 square pixels of 3.5 µm dimension. The system was pressure-tested and certified to operate to a depth of 1.8 km (18 MPa water pressure). The light path in-water between the windows is 45 cm giving a recording volume of about 36 cm^3 of the water column in a single hologram.

eHoloCam was deployed in the North Sea from the RV Scotia (Marine Scotland Science, Marine Laboratory, Aberdeen) on four cruises covering 1 year (December 2005, April, July and December 2006). It was operated from a sampling frame "Auto-Recording Instrumented Environmental Sampler, Marine Scotland Science, Aberdeen" (ARIES) and towed at up to 4 knots to depths of 450 m. Several hundred

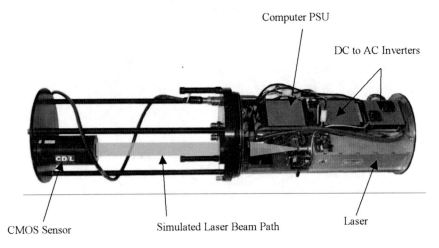

Fig. 5.13 *eHolocam* showing sensor housing (*left*), main housing (*right*) with beam path highlighted

holovideos were recorded over the complete season of dives. The CMOS camera is operable in different pixel configurations to offer control of resolution, frame rate and video-size. In high-resolution mode, the full array of 2,208 × 3,000, 3.5 μm square pixels is addressed at a 5.3 Hz frame rate giving a total recording of 255 frames in 48 s. In medium-resolution mode, the effective pixel size is 7.0 μm (1,500 × 1,104 pixels) at 17.3 Hz over 1,172 frames in 68 s. In low-resolution mode, the effective pixel size is 10.5 μm square (1,000 × 736 pixels) at 24.3 Hz in 110 s with 2,627 frames captured. The recording durations were calculated from the desirable video buffer size (around 2 GB). A series of holovideos, were recorded in each dive following a sequence of high, medium and low-resolution. As many as five sequences were carried out giving up to 15 separate videos per dive.

Hologram frames were manually reconstructed using the angular spectrum algorithm (prior to the incorporation of autofocusing) and images extracted from the recorded e-holovideos. Here, we present some examples of the reconstructed

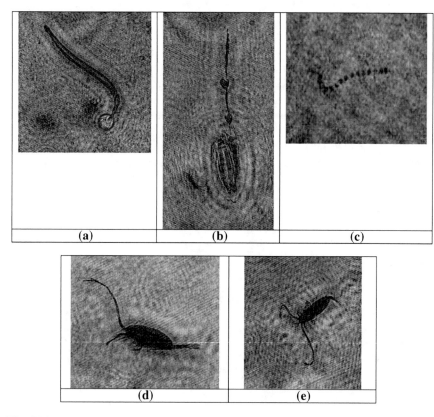

Fig. 5.14 Reconstructions (at different scales) of digital holograms recorded in the North Sea using eHoloCam; **a** chaetoganatha, about 4 mm long, **b** a fragment of jelly fish larvae, body length about 2.5 mm, **c** a phytoplankton chain, about 500 μm long, **d** and **e** calenoid copepods of about 2.5 mm body length

images (Fig. 5.14) from several dives over different seasons; the types of plankton, their population density and behavioural characteristics varied significantly between seasons and locations. Many of the species recorded are very fragile and can be damaged by most sampling methods, such as net collection. Several varieties of jellyfish larvae were observed and these demonstrate the advantage of holography in being able to record semi-transparent (phase) objects.

A particular set of ten holovideos were recorded in sequence at high, medium and low resolution and illustrate the analysis of population densities of calanoid copepods. The total sampling volume over all videos was 400,000 cm^3 and 79 copepods were identified giving an average population of 196×10^{-6} cm^3.

eHoloCam has since been redesigned and reconfigured to operate to 10,000 m depth.

Chapter 6
Special Techniques

6.1 Applications Using Short Coherence Length Light

6.1.1 Light-in-Flight Measurements

Holographic recording of Light-in-Flight (LiF) was first proposed by Abramson [1–4]. He pointed out that a hologram can only image the distances in space where the optical path of the reference wave matches that of the object wave. The basic idea of LiF consists of recording a hologram of a plane object by a short coherence length laser, Fig. 6.1. For this purpose a cw Ar-Ion laser without intracavity etalon could be used. The coherence length of such laser is in the range of few millimeters or less. Alternatively also a picosecond pulsed dye laser can be used. The reference wave is guided nearly parallel to the holographic plate (grazing incidence). In this way, only those parts of the object are recorded (and later reconstructed), for which the optical path difference (OPD) between object and reference wave is smaller than the coherence length of the light source. By changing the observation point in the developed plate, the above condition is met for different parts of the object, thus allowing observation of a wavefront as it evolves over the object.

Digital Holography has been applied to LiF recordings by Pomarico et al. [100, 189]. In this work, an Ar-Ion laser pumped cw dye laser (Rhodamine 6G, $\lambda = 574$ nm) is used. No frequency selecting elements are installed in the laser resonator. Therefore the output spectrum consists of many oscillating modes, resulting in a coherence length, which is determined by the LiF experiments to be 2.3 mm.

The laser beam is divided into a plane reference wave illuminating the CCD array and into a diverging wave illuminating the object, Fig. 6.2. The path differences are provided by glass plates with different but known thicknesses. The object consists of a plane aluminum plate of 2 cm × 2 cm area, the distance between object and CCD sensor is 1.67 m, and the angle of illumination α (referred to the normal of the object) is about 80°. A wavelength of $\lambda = 574$ nm is used and the maximum

U. Schnars et al., *Digital Holography and Wavefront Sensing*,
DOI 10.1007/978-3-662-44693-5_6

Fig. 6.1 Light-in-flight holography

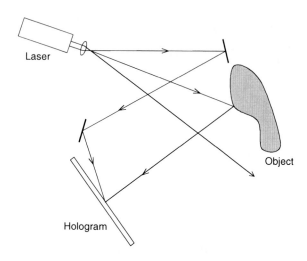

Fig. 6.2 Optical set-up for Digital LiF recording with delay lines by glass plates

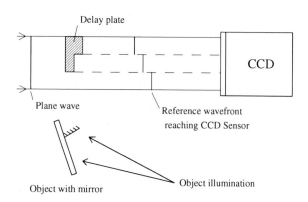

angle between object and reference wave is 2°. A camera with 2,048 × 2,048 pixels is used as recording medium.

The first experiment is performed without any glass plates in the reference arm of the interferometer. Since the object is illuminated at an angle only a part of the object wave is coherent with the reference wave. The numerical reconstruction of such a digitally recorded hologram shows, as expected, a bright stripe representing the wavefront, Fig. 6.3.

The reconstructed image is available in digital form and further processing is easy accomplished. For example, the coherence length can be calculated from this image, Fig. 6.4. The width of the bright stripe (wavefront) is determined from both, the coherence length of the light source, L, and the geometrical conditions of the holographic set-up. If a plane reference wave is used and the angle between the interfering waves is small ($\theta_{max} = 2°$ in this example), only changes in the optical path due to the illumination beam have to be considered. In this case the bright zone at the object has a width w given by

Fig. 6.3 Numerically
reconstructed wavefront

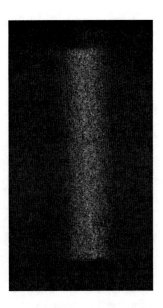

$$w = \frac{c\tau}{\sin \alpha} = \frac{L}{\sin \alpha} \tag{6.1}$$

where α is defined in Fig. 6.4 and τ is the coherence time. After measuring w,
Eq. (6.1) can be used to calculate the coherence length L of the light source using
the known angle α of the incident wave. As the measurements of the width are
disturbed by electronic and coherent noise, direct measurement of the intensity
profile leads to errors. A good result can be achieved by low-pass filtering of the
image and by applying the autocorrelation function to the intensity profile line. The
width of the wavefront measured by this procedure equates to 45 pixels. The
experimental conditions are: $\Delta x = 9$ μm; $d = 1.67$ m; $\lambda = 574$ nm; $\alpha = 80°$

Fig. 6.4 Geometrical
considerations for calculating
the coherence length

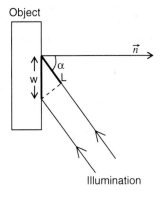

The resulting coherence length is therefore

$$L = w \sin \alpha = 45 \cdot \Delta\xi \sin \alpha = 2.3 \, \text{mm} \tag{6.2}$$

where Eq. (3.23) is used for calculating the image resolution $\Delta\xi$.

It is also possible to apply Digital Holography to follow the evolution of a wave front in its "flight" over an object, as proposed in the original work of Abramson for conventional holography. However, because of the reduced size of the CCD target and the lower resolution compared to a holographic plate, only slightly different points of view of the wave front can be registered in each hologram.

A possible setup for this purpose, using a skew reference wave, has been proposed by Pettersson et al. [187]. However, this solution is not applicable in this situation because the high spatial frequencies that would be produced at the sensor are not resolvable. A solution to this problem is to record a hologram introducing different phase delays in different parts of the plane reference wave. That can be achieved e.g. by introducing plane-parallel plates of different thickness in the plane wave illuminating the CCD sensor, as in Fig. 6.2. A plate of thickness p and refractive index n will produce a delay Δt with respect to air (or in vacuum with light speed c) given by:

$$\Delta t = (n - 1)\frac{p}{c} \tag{6.3}$$

That way it is possible to record in one exposure several holograms of the object using a corresponding number of reference waves delayed with respect to each other. The numerical reconstruction can carried out for each part of the CCD array in which the phase of the reference wave has a particular delay, giving rise to the desired "times of evolution" of the wave front illuminating the object. This is equivalent to choose another observation point in the original LiF experiment. In this sense, the phase delays introduced in the different parts of the reference wave can be interpreted as artificial extensions of the CCD sensor and allow a better visualization of the phenomenon.

In these experiments 6 mm thick PMMA plates (refractive index n $\sim$ 1.5) are used to produce the phase delays in the reference wave, Fig. 6.2. One third of the original plane reference wave does not travel through PMMA, the second third, illuminating the sensor in the middle, travels through 6 mm PMMA (representing 10 ps delay with respect to air) and the last third travels through 18 mm PMMA (30 ps delay with respect to air). The object as seen from the CCD sensor is schematically sketched for better recognition of the results, Fig. 6.5. It consists of a 3 cm × 3 cm plane aluminum plate, which was painted matt white for better light scattering. A small plane mirror (1 cm × 1 cm area) is attached to the plate, perpendicular to its surface and at an angle of about 10° to the vertical.

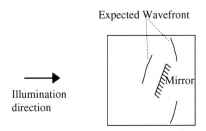

Expected Wavefront

Illumination
direction

Mirror

Fig. 6.5 Object used for displaying the temporal evolution of a wave front as seen from the CCD sensor

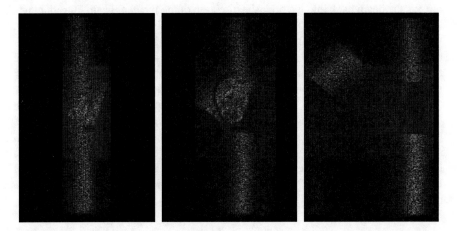

Fig. 6.6 The wavefront at three different times, reconstructed from one single holographic recording. *Left* no delay, wavefront just reaching mirror. *Middle* 10 ps delay, the mirror reflects one part of the wavefront. *Right* 30 ps delay with respect to the *left* recording, one part is reflected into the opposite direction, the other part is traveling in the original direction

The three reconstructed stripes of the hologram, corresponding to three different times of evolution of the wavefront illuminating the object are shown in Fig. 6.6. One part of the wavefront is reflected by the mirror, the other part is traveling in the original direction. The three pictures can be interpreted as a slow-motion shot of the wavefront. As demonstrated before, quantitative results can be derived from these images, e.g. the speed of light.

The minimum number of pixels required for a part of the hologram to be successfully reconstructed limits the number of different "times of evolution" that can be simultaneously recorded. Furthermore, due to the borders of the plates introduced into the reference wave, diffraction effects cause dark zones at the CCD which cannot be used for numerical reconstruction.

6.1.2 Short-Coherence Tomography

The main disadvantage of introducing the path differences by glass plates with
different thickness are the diffraction effects at the edges of the plates. Therefore LiF
only allows a few discrete depth steps to be incorporated. To overcome this problem,
Nilsson and Carlsson proposed the use of a blazed reflection grating for generating
path differences [25, 162–164]. The set-up is composed of a Michelson Interferometer, Fig. 6.7, in which one mirror is replaced by the grating. The incoming beam
of a light source with sufficient short coherence length is split into two partial beams.
One partial beam illuminates the object and is diffusely reflected from the surface to
the CCD. The other beam is guided to the blazed reflection grating. The grating
reflects the beam back into the opposite direction of the incident beam, introducing a
spatially varying delay across the beam profile. Both beams interfere at the CCD,
which records the hologram. The method can be applied to measure the three-
dimensional object shape. This is possible because each vertical stripe of the
hologram fulfils the coherence condition for different object depths. Reconstruction
from different hologram parts creates different depth layers of the object.

Instead of the grating in Fig. 6.7 it is also possible to use an ordinary mirror in
the reference arm, see e.g. [179], which can be shifted in the direction of beam
propagation. Digital holograms are then recorded in different mirror positions. Each
single hologram represents another depth layer of the object and the shape can be
calculated from the holographic reconstructions. However, there is an advantage of
the setup shown in Fig. 6.7 using the grating: only one recording is necessary to
determine the whole object shape.

Fig. 6.7 Short-coherence
length tomography using a
blazed grating

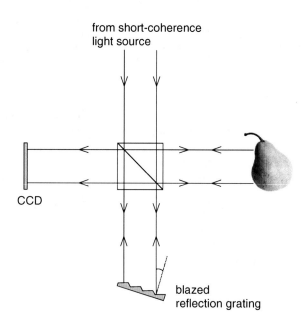

6.2 Endoscopic Digital Holography

Digital Holography provides the possibility to combine deformation measurement and surface contouring in one single set-up. In the simplest case of measuring object deformation, two holograms with a known wavelength are recorded for different object states (Sect. 4.2). When shape measurement is required, the object has to remain unchanged while two holograms with slightly different wavelengths or slightly different illumination points are recorded (Sect. 4.3). Thus, and due to the relatively simple geometry, this method appears to be well suited to endoscopic measurements [101, 119] and the following description is partly based on these publications. The requirements for an endoscopic Digital Holography sensor are much more stringent than they are for a laboratory breadboard; for example, an endoscopic system has to,

- be more flexible;
- be more robust in harsh environments;
- incorporate faster data processing;
- be very small;
- be adapted to restrictions caused by the system size.

A sketch of a prototype system is shown in Fig. 6.8, while Fig. 6.9 depicts a functional prototype of the sensor head. The system can be divided into four parts: the controlling computer, the laser and the corresponding fibre coupling units, the endoscope and the sensor.

The sensor head has a diameter of 15 mm (current stage). In future it is intended to decrease the size to a diameter of less than 10 mm.

The heart of the sensor is a commercial miniature CCD-camera with a 1/3″CCD-matrix. Including the housing, this camera has a diameter of 10 mm. The objective of the camera is removed to be able to record the digital holograms. Since the camera provides a standard video-signal the hologram can be grabbed by a conventional frame grabber. For the object- and the reference beam mono-mode glass fibres are used. Currently, a single illumination beam is utilized. This is sufficient to measure

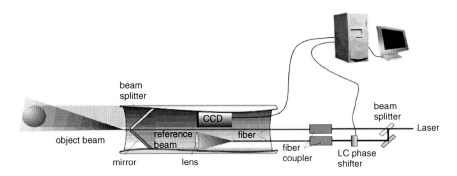

Fig. 6.8 Sketch of an endoscope based on digital holography

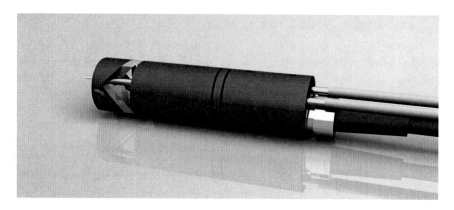

Fig. 6.9 Sensor head of the endoscope (*photo* BIAS)

the shape of the object and to measure one displacement component—in our case the out-of-plane component. However, in the next step three illumination directions will be implemented to be able to perform a 3D displacement measurement.

In general every laser source with sufficient coherence length can be used for deformation measurements. However, for shape measurements, the wavelength of the laser needs to be tuneable over a suitable spectral range. Thus, to keep the whole system portable, currently a VCSEL laser diode is used as a light source. The wavelength for this laser diode type can be tuned continuously in a range of about 8 nm.

From Fig. 6.8 it can be seen that the laser beam passes through a liquid crystal phase shifter before it is coupled into the fibre for the reference beam. This LC phase shifter is used to record temporal phase shifted holograms. A simple reconstruction without using phase shifting results in an image that contains the desired object image and additionally the twin image together with the zero order term. By using temporal phase shifting the conjugate image as well as the zero order can be eliminated completely from the resulting image (see Sect. 3.3.3). In this way the full space-bandwidth of the CCD can be utilized. This is of great importance, since the choice of the camera is restricted by the system size. Cameras of this size are only available with a limited pixel number, which makes it necessary to make use of all available pixels.

The high sensitivity of Digital Holography to object motion is also a disadvantage for a system that is intended to be used outside the laboratory. Even small object vibrations caused by environmental influences can disturb the measurement, and so high processing speed and fast data acquisition are important to minimize the influence of unwanted disturbances. In order to achieve a high processing speed an optimized phase-shift algorithm has been chosen [138]. More than six reconstructions per second are possible for holograms with 512×512 pixels using a PC with 1.4 GHz clock frequency.

Another benefit of high processing speed is the possibility to unwrap the phase maps of deformation measurements in real time by temporal phase unwrapping [91].

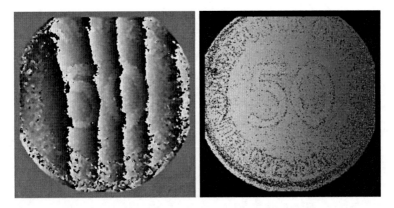

Fig. 6.10 Wrapped deformation phase of a heat loaded coin (*left*) and unwrapped phase generated by temporal unwrapping (*right*)

In this method the total object deformation is subdivided in many measurement steps in which the phase differences are smaller than 2π. By adding up those intermediate results, the total phase change can be obtained without any further unwrapping. This is an important feature, since it is essential to have an unwrapped phase to be able to calculate the real deformation data from the phase map. Figure 6.10 shows an example of such a measurement. The left image shows the wrapped deformation phase for a clamped coin which was loaded by heat. The right image shows the temporal unwrapped phase, which has been obtained by dividing the total deformation in 85 sub-measurements.

6.3 Optical Reconstruction of Digital Holograms

The techniques discussed in this chapter differ from some of the other methods presented previously, since reconstruction is performed optically. The computer is just used as intermediate storage medium for digital holograms and some means of performing optical read-out is needed, see Kujawinska et al. [133, 269].

Liquid Crystal Displays (LCD's) are electro-optic devices used to modulate light beams and they can be used as a spatial light modulator (SLM) in holography. An individual LCD cell changes its transmittance depending on the applied voltage. It is therefore possible, by varying the voltage, to modulate the brightness of light, which passes through the device.

Optical hologram reconstruction with a LCD is possible e.g. with the set-up of Fig. 6.11. Firstly, a hologram is recorded on an electronic sensor, Fig. 6.11a. The hologram is stored and then transmitted to the reconstruction set-up, Fig. 6.11b. Recording and reconstruction set-ups could be located at different sites. The LCD modulates the reconstruction beam with the hologram function. The original object wave is reconstructed due to the diffraction of the reconstruction beam at the

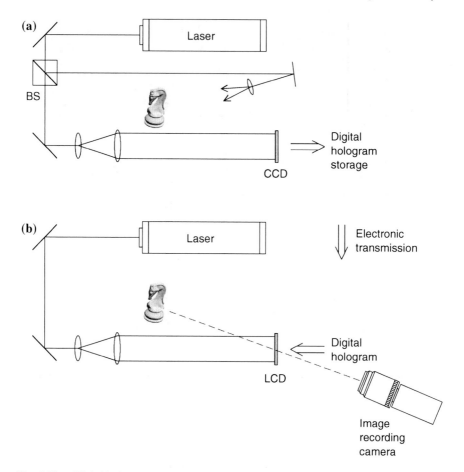

Fig. 6.11 a Digital hologram recording with a CCD. **b** Optical reconstruction with a LCD

modulated LCD. The virtual image can be observed at the position of the original object. Alternatively it is possible to reconstruct the real image by illuminating the LCD with the conjugate of the reference wave.

An example of such an optical reconstruction of a digitally recorded hologram is shown in Fig. 6.12a. A digital hologram of a chess piece knight is recorded and stored. The image of the knight becomes visible if the LCD with the hologram mask is illuminated by a reconstruction wave. Optical reconstruction of two superimposed holograms, which are recorded in different object states results in a holographic interferogram, Fig. 6.12b.

Instead of an LCD other electro-optical devices can be used as spatial light modulators, too. Kreis, Aswendt und Höfling published the optical reconstruction by means of a Digital Mirror Device (DMD) [128]. A DMD is a silicon micromachined component, which consists of an array of tiltable aluminium mirrors mounted on hinges over a CMOS static random access memory (SRAM). Today

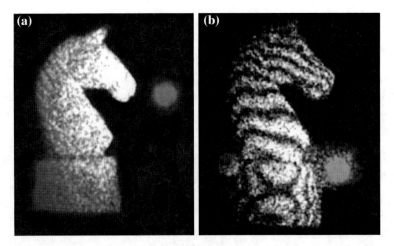

Fig. 6.12 a Optical reconstruction of a digital hologram by means of a LCD. **b** Optically reconstructed holographic interferogram (from [173])

DMD's are available with up to 1280 × 1024 mirror elements. The individually addressable mirrors can be tilted in binary mode either −10° (on) or +10° (off) along an axis diagonal to the micromirror. In optical reconstruction DMD's are therefore operated in reflection. In contrast to LCD's, which absorb up to 90 % of the available light, a DMD is a reflective device yielding much more light. Consequently the diffraction efficiency in the hologram reconstruction is better when compared with LCD's.

A very simple device for displaying digital holograms is a computer printer. The high resolution of standard ink-jet or laser printers with up to 3,000 dots per inch makes it possible to print digital holograms directly on a transparent film. The hologram is then reconstructed by illuminating this film with the reconstruction wave.

6.4 Comparative Digital Holography

6.4.1 Fundamentals of Comparative Holography

The principle of interferometry is the comparison of the optical wave reflected or transmitted by the test object with another, known wavefield [102]. In Holographic Interferometry at least one of these waves is stored by a hologram. By interference the phase difference between the two wavefields can be measured. The phase differences are related to the quantities to be determined via the geometry function of the set-up. In this way it is possible to measure object shapes or deformation. However, a severe restriction in conventional HI is that interference is only possible if the microstructures of the surfaces to be compared are identical. The replacement of the object or large deformations lead to a decorrelation of the two speckle fields and the loss of the interference. Thus standard HI is restricted to the comparison of two states of the *same* object.

A method to overcome this restriction is comparative interferometry [67, 161]. This method is based on the illumination of the two states of the test component with the corresponding conjugate object wave of the master object: the object wave of the master component acts as a coherence mask for the adaptive illumination of the test component.

In comparative interferometry a double exposure hologram of the master object is taken in the two states according to a specific load. Reconstruction of this double exposed hologram generates an interferogram. The relation between the measured interference phase and the displacement vector is given by Eq. (2.84):

$$\Delta\varphi_1(x,y) = \frac{2\pi}{\lambda}\vec{d}_1(x,y,z)\left(\vec{b}_1 - \vec{s}_1\right) = \vec{d}_1\vec{S}_1 \qquad (6.4)$$

The test object is investigated in a modified set-up for comparison with the master: it is illuminated in the original observation direction $\vec{b}_1$ by the reconstructed, conjugate wavefront of the master object, i.e. the real image of the master object is projected onto the test object. It is observed in the original illumination direction $\vec{s}_1$. This procedure results in

$$\vec{s}_2 = -\vec{b}_1 \quad \text{and} \quad \vec{b}_2 = -\vec{s}_1 \qquad (6.5)$$

$$\Delta\varphi_2(x,y) = \frac{2\pi}{\lambda}\vec{d}_2(x,y,z)\left(\vec{b}_2 - \vec{s}_2\right) = \frac{2\pi}{\lambda}\vec{d}_2(x,y,z)\left(\vec{b}_1 - \vec{s}_1\right) \qquad (6.6)$$

Since the test object is illuminated by the conjugated wavefront of the master the interferogram indicates the difference of the displacements between the two objects:

$$\Delta\varphi(x,y) = \Delta\varphi_1(x,y) - \Delta\varphi_2(x,y) = \frac{2\pi}{\lambda}\left(\vec{d}_1(x,y,z) - \vec{d}_2(x,y,z)\right)\left(\vec{b}_1 - \vec{s}_1\right) \qquad (6.7)$$

6.4.2 Comparative Digital Holography

Comparative Digital Holography is a combination of comparative holography with Digital Holography [173, 174]. A digital hologram of a master object is recorded at a location A, Fig. 6.13a. The transmission of this digital hologram to a test location B can be done by any data transfer medium, e.g. by the internet. At location B the hologram is fed into a Liquid Crystal Display operating as a spatial light modulator. A laser reconstructs the hologram optically.

For the comparative holography the conjugate wavefronts of the master object are reconstructed and illuminate the test object, Fig. 6.13b. The observation is done in the original illumination direction. A great advantage of comparative DH

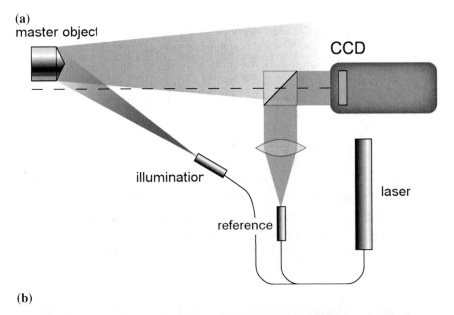

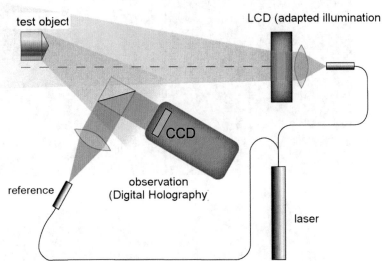

Fig. 6.13 Comparative digital holography (from [173]). **a** Recording of the mask. **b** Coherent illumination of the test object with the conjugated wavefront of the master

compared to conventional comparative HI is, that the holograms of all states can be stored and later reconstructed separately from each other. Therefore no additional reference waves are needed for the separate coding of the different holograms. This characteristic of Digital Holography reduces the technical requirements for comparative measurements significantly.

The method is demonstrated by the determination of a small dent in one of two macroscopically identical cylinders with cones at their upper end. The depth of the dent is a few micrometers. With holographic two-wavelength contouring, the observed phase differences can be described by

$$\Delta\varphi_1(x,y) = \frac{2\pi}{\Lambda}\left(\vec{b}_1 - \vec{s}_1\right)\overrightarrow{\Delta r}_1(x,y) \tag{6.8}$$

$$\Delta\varphi_2(x,y) = \frac{2\pi}{\Lambda}\left(\vec{b}_2 - \vec{s}_2\right)\overrightarrow{\Delta r}_2(x,y) \tag{6.9}$$

The indices 1 and 2 denote the master or the test object, respectively, Λ is the synthetic wavelength. The measurements are carried out with $\Lambda = 0.345$ mm ($\lambda_1 = 584.12$ nm and $\lambda_2 = 585.11$ nm), $\overrightarrow{\Delta r}_1$ and $\overrightarrow{\Delta r}_2$ represent the relative height deflection of the master with respect to the test object. Figure 6.14a shows the reconstructed intensity of the test object, while the mod 2π contour lines are depicted in Fig. 6.14b. The damage site is hard to recognize. However, after

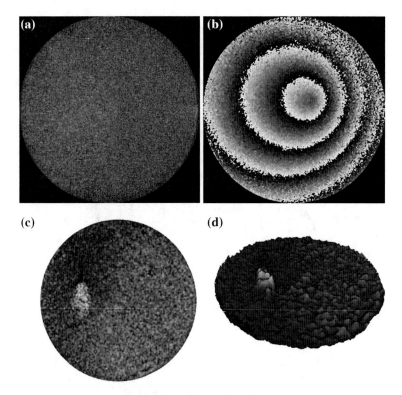

Fig. 6.14 Demonstration of comparative digital holography (from [174]). **a** Reconstructed intensity, test object. **b** Phase contour lines, test object. **c** Comparative phase difference, mod2π-map. **d** Comparative phase difference, pseudo 3D-map

holographic illumination of the test object with the real image of the master, the difference phase $\Delta\varphi$ corresponds to the difference in height deflections between master and test object:

$$\Delta\varphi(x,y) = \Delta\varphi_1(x,y) - \Delta\varphi_2(x,y) = \frac{2\pi}{\Lambda}\left(\overrightarrow{\Delta r}_1(x,y) - \overrightarrow{\Delta r}_2(x,y)\right)\left(\vec{b} - \vec{s}\right)$$

$$(6.10)$$

This phase difference is shown in Fig. 6.14c (mod 2π-map) and Fig. 6.18d (pseudo 3D-map).

The measured phase difference distribution is quite noisy because of the large pixel dimensions of the CCD target and the spatial light modulator (CCD: 9 μm, LCD: 29 μm). In future, better optical components might be available. Nevertheless, the comparison of Fig. 6.14b with 6.14d demonstrates the advantage of comparative Digital Holography to measure the shape difference of two objects with different microstructure: in the phase difference image the dent is clearly recognizable.

6.5 Encrypting of Information with Digital Holography

Reconstruction of objects from their holograms is only possible, if the reconstruction wave has nearly the same properties as the original reference wave used in recording. Any deviation from the original amplitude and phase distribution results in image aberrations or in total loss of the original object information. The recording reference wave can be therefore regarded as a key to reconstructing the information coded in the hologram. This is the principle of information encryption by holography.

In the following a coding method proposed by Javidi et al. [95, 231, 232] is described. The method is based on phase-shifting Digital Holography, see set-up in Fig. 6.15. The key for encrypting the information is a diffusely scattering screen. A parallel beam is split into two coherent partial beams at beam splitter BS1. One partial beam illuminates the screen from the back.

The scattered light is guided to the CCD via beam splitter BS2. The other beam is guided via BS3, mirror M3 and BS2 to the CCD. For this shutter SH1 is opened, shutter SH2 is closed and the object is removed. Both beams interfere at the surface of the CCD. A set of four interferograms with mutual phase shifts is recorded by means of phase shifting devices. This can be done either by aligning the fast and the slow axes of optical retarders with the polarization of the incident beam (as shown in Fig. 6.15) or by other electro-optical devices like piezo-electric driven mirrors. The complex amplitude of the plane partial wave guided via mirror M3 is $1 \cdot e^{i \cdot 0}$ in the simplest case. Consequently, it is possible to calculate the complex amplitude $a_K e^{i\varphi_K}$ of the wave scattered from the diffuser in the CCD plane by phase shifting

Fig. 6.15 Information
encrypting with digital
holography

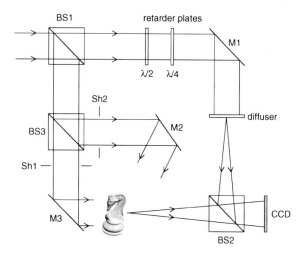

Fig. 6.15 Information encrypting with digital holography

algorithms (see also Sect. 2.7.5). If four interferograms I_1–I_4 with equidistant phase steps of $\pi/2$ are recorded the amplitude and the phase of the "key" wave are determined by following equations:

$$\varphi_K = \arctan \frac{I_4 - I_2}{I_1 - I_3} \tag{6.11}$$

$$a_K = \frac{1}{4} \sqrt{(I_1 - I_3)^2 + (I_4 - I_2)^2} \tag{6.12}$$

Recording of the object to be encrypted is accomplished by closing shutter SH1 opening shutter SH2 and illuminating the object via M2. The scattered light from the screen is now used as reference wave. Again a set of four phase shifted interferograms I'_1–I'_4 is generated. The phase difference between the "key" wave phase φ_K and the object phase φ_0 is determined by:

$$\varphi_0 - \varphi_K = \arctan \frac{I'_4 - I'_2}{I'_1 - I'_3} \tag{6.13}$$

The following equation is valid for the product of the amplitudes:

$$a_0 \cdot a_K = \frac{1}{4} \sqrt{(I'_1 - I'_3)^2 + (I'_4 - I'_2)^2} \tag{6.14}$$

Without knowledge of a_k and φ_K it is obviously not possible to calculate the complex amplitude of the object wave in the CCD plane. The object can only be reconstructed with the numerical methods described in Chap. 3, if the correct key is given. This key consists of amplitude and phase of the wave scattered from the diffuser. A second key, which has to be known for correct object decoding, too, is the recording distance d between diffuser and CCD.

6.6 Synthetic Aperture Holography

Every part of an off-axis hologram encodes the entire information about object. The object can be reconstructed therefore from any cut-out of the hologram, where the size of such a cut-out only influences the speckle size in the reconstructed image. On the other hand it is also possible to synthesize a hologram from different single holograms [129]. A possible recording geometry with two CCD's is depicted in Fig. 6.16 for a plane reference wave. Independent CCD-cameras are used to record the single holograms. A fixed phase relation between the individual holograms is ensured by using the same reference wave. Reconstruction of the synthesized hologram is possible by following methods: single holograms are reconstructed separately and the resulting complex amplitudes in the image plane are coherently superimposed. A second possibility is to embed both single holograms in an artificial large hologram, where the grey values of all pixels not covered are set to zero (black). Such an artificial hologram matrix is then reconstructed as a whole.

   The resolution of images and phase maps reconstructed from digital holograms depends on the recording distance d and on the effective aperture $N\Delta x$, see Eq. (3.23). However, both quantities are not independent of each other, because for state-of-the-art sensors with pixel sizes in the range of 5 μm large apertures require also long recording distances due to the spatial frequency limitations discussed in Sect. 3.4.2. Increasing the aperture size by using more than one sensor therefore does not automatically improve the image resolution, because the larger synthetic aperture requires a longer recording distance. In order to decouple recording distance and aperture size it is therefore necessary to use sensors with small pixel sizes

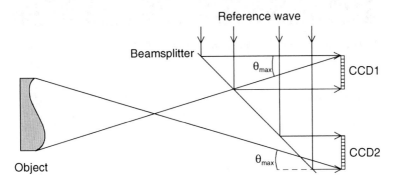

Fig. 6.16 Aperture synthesis with two CCD's

in the range of one micron or below, which might be available in the future. With such devices even the highest spatial frequencies could be resolved, independently from the recording distance.

6.7 Holographic Pinhole Camera

In Sect. 3.2.1 it was demonstrated, that reduction of the effective aperture size leads to a reduction of the image resolution. In Fig. 6.17 (left) the die used in Chap. 3 is reconstructed using only 255 × 255 pixels, i.e. only one sixteenth of the original hologram area. The die is still visible but with larger speckle size. Further reduction of the aperture size to only 128 × 128 effective pixels (1/64 of the original hologram matrix) increases the speckle size further so that the object can only be recognized with very low resolution, see Fig. 6.17 (right). However, on the opposite side of the DC term the other image, which was previously totally defocused, also becomes visible in low resolution. The reason for this double image is due to the "pinhole camera effect". Light from the out-of-focus virtual image passes through the small aperture (pinhole) and projects an inverted image on the opposite side of the CCD.

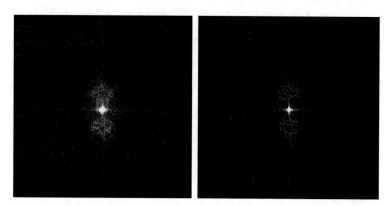

Fig. 6.17 *Left* reconstruction with 256 × 256 effective pixels. *Right* reconstruction with 128 × 128 effective pixels

Fig. 6.18 The pinhole camera effect

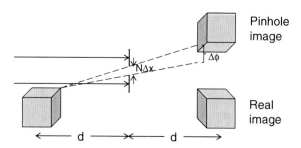

The smaller the aperture size, the higher the resolution of the pinhole camera image becomes, see Fig. 6.18. In a distance d the image resolution of the pinhole camera image is given by the projected circle of confusion (can be seen from Fig. 6.18):

$$\Delta\phi = 2N\Delta x \qquad\qquad (6.15)$$

That means holographic image forming and image generation due to the pinhole effect are opposing effects. A smaller pinhole (small hologram aperture) will result in sharper image resolution of the pinhole image, while a large aperture is needed to create high resolution holographic images.

Chapter 7
Computational Wavefield Sensing

7.1 Overview

With the advent of faster computer processors, alternative methods of wavefield sensing have been developed throughout the past decades. In contrast to standard interferometry, these methods aim at solving an inverse problem, whereby the recorded intensities are interpreted as an effect caused by the underlying (unknown) wavefield when subjected to different manipulations. In contrast to holography and interferometry it is not possible to use film as a recording material and to optically reconstruct the wavefield. In fact, it is even pertinent to say that the numerical task of solving the inverse problem is an essential and integral part of the measurement process. Bearing this in mind we may categorise these techniques under the term *computational wavefield sensing*, in analogy to similar efforts in the field of imaging.

In most cases reasoning back on the wavefield subject to a set of observed intensities is both mathematically and computationally demanding, calling for sophisticated numerical methods and evaluation procedures. Having said this, there are great benefits offered by computational wavefield sensing. It provides the means to determine the complex amplitude of a wavefield without the requirement of a particular reference wave. In optical metrology, this enables measurements with interferometric accuracy and precision but yet strongly reduced demands with respect to temporal and spatial coherence of the light as well as the mechanical stability of the environment. It allows interferometry with low brilliance (number of photons per time, area and solid angle within a small spectral range) light sources, such as light emitting diodes (LED) with large emitters and even liquid-crystal displays (LCD). This option is eye-safe, cheap and significantly reduces the disturbing effect of coherent amplification arising from parasitic reflections within the optical setup. Finally, it allows for determination of wavefields in cases in which the application of a reference wave is impossible, such as in stellar interferometry.

The above properties are the main reasons why computational methods have created substantial interest over the past decade or so. A comprehensive discussion

© Springer-Verlag Berlin Heidelberg 2015
U. Schnars et al., *Digital Holography and Wavefront Sensing*,
DOI 10.1007/978-3-662-44693-5_7

and survey of the field including all existing techniques is beyond the scope of this chapter. The aim of the following sections is therefore to provide an overview over the most prominent methods, such as iterative and deterministic phase retrieval, computational shear interferometry (CoSI) and Hartmann-Shack wavefront sensing.

7.2 Phase Retrieval

Initially the name *Phase Retrieval* referred to the problem of reconstructing an image from the modulus of its Fourier transform. Obviously recovering the phase in the Fourier domain from some pre-knowledge or constraints would reduce the problem to a simple inverse Fourier transform. Similarly, in coherent optics, phase retrieval refers to techniques in which the intensity of a wavefield is known across one or several separated planes perpendicular to the axis of propagation. The task then is to find a set of corresponding phase distributions which is consistent with the Helmholtz-Equation and, if available, some additional constraints. In this section we will present the mathematical foundations of the phase retrieval problem in coherent optics and some of the most established ways to solve it.

We will begin with defining a notation for the above statement. Let us consider a wavefield propagating through free space along the z-axis. Let us further define a sequence of N planes at positions $\{z_n | n = 1 \dots N\}$ perpendicular to z as seen from Fig. 7.1. We will refer to the complex amplitude of the wavefield across the nth plane as

$$E_n(x, y) = A_n(x, y) \cdot \exp[i\theta_n(x, y)]. \tag{7.1}$$

By construction, all E_n satisfy the Helmholtz-Equation, which means that two complex amplitudes E_n and E_m can be related by any propagation operator P, such that

Fig. 7.1 Phase retrieval is based on measuring the intensity of a wavefield across a number of parallel planes along the main axis of propagation

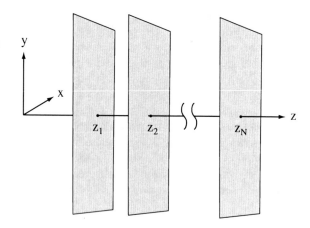

$$E_m(x, y) = P\{E_n; z_m - z_n\}. \tag{7.2}$$

where $z_m - z_n$ is the propagation distance between the planes. An example of an explicit implementation of the propagation operator is given by the Fresnel-Kirchoff diffraction integral in Eq. (2.48). We can assume the intensities $I_n(x, y) = E_n(x, y)E_n^*(x, y)$ of the wavefield across all of the planes known but we do not know the true phase distributions $\theta_n(x,y)$. With this notation we can formulate the inverse problem constituting the phase retrieval scheme by means of the following objective function in a least-squares sense:

$$
\begin{aligned}
L(f_1) &= \sum_n \left\| |P\{f_1; z_n - z_1\}| - \sqrt{I_n(x, y)} \right\|^2 \\
&= \sum_n \iint \left| |P\{f_1; z_n - z_1\}| - \sqrt{I_n(x, y)} \right|^2 dxdy,
\end{aligned}
\tag{7.3}
$$

where the L^2-norm $\|...\|^2$ indicates integration of the absolute squares over all positions (x,y). In the discrete case the integral in Eq. (7.3) is replaced by a sum. The function $f_1(x,y) = A_1(x,y)\exp[i\varphi_1(x,y)]$ is an estimate of the complex amplitude $E_1(x,y)$ of which only the amplitude $A_1(x,y)$ is known from the square root of the measured intensity. Solution of the phase retrieval problem requires minimization of the functional L with respect to f_1 (or sometimes directly φ_1) hence the solution is given by the best estimate

$$\tilde{E}_1(x, y) = \min_{f_1} L. \tag{7.4}$$

Note that because of the propagation operator, Eq. (7.3) is inherently consistent with the Helmholtz-Equation. Once the wavefield in the first plane is estimated, the phase can be determined in any of the planes by

$$\tilde{\varphi}_n(x, y) = \arg\{P\{\tilde{E}_1; z_n - z_1\}\}. \tag{7.5}$$

A large number of different approaches have been developed to attempt the minimization implied by Eq. (7.4). They can be mainly distinguished into iterative and direct methods which we will discuss separately in the next sections. Some techniques are not even designed to find an optimum in the strict least-squares sense suggested by Eq. (7.3) but are computationally very efficient and still yield acceptable results.

The choice of the minimization technique as well as the number of measurements strongly influences the accuracy of the phase estimation. Additionally, the results can be largely affected by including pre-knowledge into the minimization process, such as limited support either in the spatial or the spectral domain or non-negativity for example. However, by construction all phase retrieval techniques strongly rely on intensity diversity, i.e. that the measured intensities change significantly between the observation planes.

7.2.1 Projection Based Methods

Probably one of the most established iterative methods has been published by Gerchberg and Saxton [71] in 1972 and generalized for application in the field of electron microscopy by Misell [158] in 1973. It is based on two intensity observations of the same wavefield in two propagation states. While in the initial Gerchberg-Saxton approach the propagation states correspond to a spatial domain and the corresponding Fourier domain, i.e. the far field, Misell only required a minimum of two defocused representations of an electron beam to retrieve phase values even in the case of partial coherence. The basic scheme of these techniques is seen from Fig. 7.2. Both start with assuming an initial estimate of the phase function in the first plane $g_1 = \exp[i\varphi_1]$ with random phase values $\varphi_1(x,y)$ in the range of $-\pi$ to π. Then the following steps are repeated until no significant changes are observed:

1. The phase function g_1 is multiplied with the amplitude obtained from the square root of the measured intensity in the first plane. This yields an estimate of the complex amplitude f_1 in the first plane

$$f_1(x,y) = \sqrt{I_1(x,y)} \cdot \exp[i\varphi_1(x,y)]. \tag{7.6}$$

2. This estimate is numerically propagated to the second plane giving the intermediate result $f_2'(u,v) = P\{f_1; z_2 - z_1\}$, where in case of the Gerchberg-Saxton

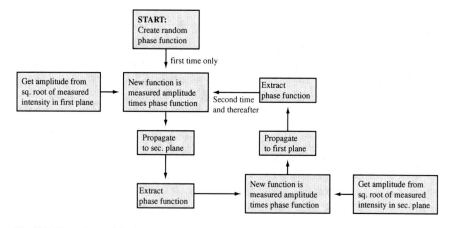

Fig. 7.2 Flow chart of the phase retrieval algorithm introduced by Gerchberg and Saxton [71]

algorithm the propagation operator P is exchanged by the Fourier transform. From this, the phase function g_2 is extracted by

$$g_2(u,v) = f_2'(u,v)/|f_2'(u,v)| = \exp[i\varphi_2(u,v)], \qquad (7.7)$$

3. which is multiplied with the amplitude known from the square root of the measured intensity in the second plane to arrive at an estimate for the complex amplitude

$$f_2(u,v) = \sqrt{I_2(u,v)} \cdot \exp[i\varphi_2(u,v)]. \qquad (7.8)$$

4. In the last step, f_2 is propagated back to the first plane to obtain the intermediate result $f_1'(x,y) = P\{f_2; z_1 - z_2\}$ and to determine a new estimate

$$g_1(x,y) = f_1'(x,y)/|f_1'(x,y)| = \exp[i\varphi_1(x,y)]. \qquad (7.9)$$

A sequence comprising all four steps is considered a single iteration of the scheme. For their approach, Gerchberg and Saxton were able to prove that the difference between the measured and the calculated amplitudes [see Eq. (7.3)]

$$\varepsilon(k) = \left\| \left| f_1^{(k)}(x,y) \right| - \sqrt{I_1(x,y)} \right\|^2, \qquad (7.10)$$

where k denotes the number of iterations and $f_1^{(k)}$ is the kth iterate, does not increase through the iterations. However they admit that the process might stagnate and that in many cases the solutions are not unique [71]. The process is usually stopped when no significant changes for $\varepsilon(k)$ are observed for consecutive iterations.

Later, Levi and Stark [143] were looking at the above algorithm from the perspective of *generalized projections* onto non-convex sets. They were able to refine the above statement and show that the method indeed converges towards a solution associated with the so called *minimum set distance*. Using their formalism it is also straight forward to generalize this statement for the defocusing method. Because it is based on a very instructive concept for the description of all Gerchberg-Saxton type approaches we will follow their argument a bit further.

The basic idea is to categorize functions sharing the same mathematical properties into *sets* in Hilbert space. For example we can define a set C_+, of which all members are functions with positive values, such as $f_1(x) = |x|$ for example. Similarly we may define a set C_A of all complex functions $f_n(x) = a(x)\exp[i\varphi_n(x)]$ which exhibit the same amplitude $a(x)$, i.e. members of C_A only differ by the phase $\varphi_n(x)$.

Sets can have the property of convexity which is best understood from Fig. 7.3. A set C is convex if the straight line segment in Hilbert space connecting any two members f_1 and f_2 of C lies entirely in C. Formally, convexity requires for all α in the range from 0 to 1

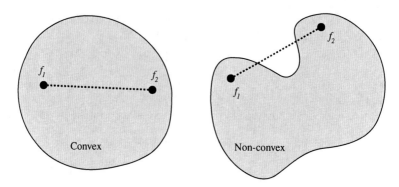

Fig. 7.3 Example of a convex (*left*) and of a non-convex (*right*) set of functions in Hilbert space. A set is called convex if a straight line between any two members of the set lies entirely in the set as well

$$f_\alpha = (1 - \alpha)f_1 + \alpha f_2 \in C \qquad (7.11)$$

Given that definition we realize that the above example of C_+ is a convex set, while C_A is non-convex. A useful tool is the projection of an arbitrary function g onto a given set C. We call $f = P_C\{g\}$ the projection of g onto C if f is the member of C that is closest to g

$$\|g - f\| = \min\|g - y\|, \quad \text{over all } y \in C. \qquad (7.12)$$

Now we can express the Gerchberg-Saxton scheme very elegantly as alternating projections onto two non-convex sets C_1 and C_2. For this, similar to the example of C_A given above, C_1 shall be the set of functions having the amplitude $a_1(x,y)$ which equals the square root of the intensity $I_1(x,y)$. Likewise, C_2 shall be the set of functions having the amplitude $a_2(x,y)$ in the diffracted plane, which equals the square root of the intensity $I_2(x,y)$. In the complex plane it is easy to verify that exchanging the amplitude of any arbitrary complex function g by a_1 while preserving the phase function of g indeed yields the projection of g onto C_1 in the sense of Eq. (7.12). By employing Parseval's theorem we get a similar statement for C_2. Hence, one iteration of the Gerchberg-Saxton algorithm can be expressed by means of two subsequent projections

$$f_1^{(k+1)} = P_1\{P_2\{f_1^{(k)}\}\}, \qquad (7.13)$$

where P_1 and P_2 denote projection onto C_1 and C_2 respectively. A good criteria to verify whether a given iterate $f_1^{(k)}$ is in agreement with the constraints defined by the sets is the so called summed distance error (SDE)

$$J\left(f_1^{(k)}\right) = \left\| P_1\left\{f_1^{(k)}\right\} - f_1^{(k)} \right\| + \left\| P_2\left\{f_1^{(k)}\right\} - f_1^{(k)} \right\|, \tag{7.14}$$

which is the sum of the distances of the iteration to the two sets. While it is proven that for convex sets alternating projections always converge to a global minimum in the sense of Eq. (7.14) [262], the situation is different for non-convex sets, yet Levi and Stark were able to prove that for consecutive iterations of Eq. (7.13) it is at least true that

$$J\left(f_1^{(k+1)}\right) \leq J\left(f_1^{(k)}\right). \tag{7.15}$$

Hence the Gerchberg-Saxton algorithm either converges or stagnates if it has found a (potentially local) minimum of the summed distance error. However, besides the fact that it often converges to a local minimum, i.e. a wrong solution, it also has a comparably slow convergence rate in many cases [58]. Therefore, throughout the past decades a number of related methods have been developed [260] by considering intensities recorded across planes connected by either a *Fresnel transform* [193] or a *fractional Fourier transform* [45]. In an attempt to provide faster converging methods, the basic scheme of numerically propagating the wavefield between two planes has been extended to multiple recording planes either by a gradient descent approach [93] or simply by extending the iterative scheme shown above to a number of consecutive planes [185]. Indeed, the method appears to converge more rapidly with an increasing number of planes involved [9]. However, to the best of our knowledge there is no proof reported that this process will yield a global minimum of Eq. (7.3), even though it is observed to find very good solutions presumably close to the minimum.

For the experimental implementation of phase retrieval it is possible to shift the camera sensor along the optical axis. However, mechanical alignment is tedious and time consuming, especially because the model introduced by Eq. (7.3) assumes the direction of the movement to exactly coincide with the axis of propagation. The setup shown in Fig. 7.4 has proven to be a convenient arrangement to experimentally realize phase retrieval in the sense that it enables recording of the required intensities in a short time and without any mechanical alignment [53].

It is composed of a classical 4f-setup with a liquid crystal spatial light modulator (SLM) in the corresponding Fourier domain as the key element. The concept makes use of the fact that the process of propagation can be expressed by a linear shift invariant system. As seen from Eqs. (3.30)–(3.33), the propagation between two planes separated by a distance z can be expressed as

$$E_2(x,y) = P\{E_1; z\} = \mathcal{F}^{-1}\{\mathcal{F}\{E_1\} \cdot h_z\}, \tag{7.16}$$

with the transfer function h_z in the spectral domain (ζ, η)

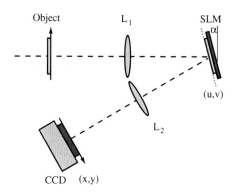

Fig. 7.4 An experimental setup for phase retrieval without any mechanically moving parts [53]. The key element is a liquid crystal spatial light modulator in the spectral domain of a 4f-setup constituted by the lenses L_1 and L_2, which modulates the incoming light with the transfer function of propagation. A tilt angle α of a few degrees is required because of the reflective SLM. It has not been observed to have any significant influence on the result and will therefore be neglected

$$h_z(\zeta, \eta) = \exp\left[ikz\sqrt{1 - \lambda^2(\zeta^2 + \eta^2)}\right], \tag{7.17}$$

and wave number $k = 2\pi/\lambda$. As seen from the structure of Eq. (7.17), propagation only affects the phase in the spectral domain, which is equivalent to the observation that the energy is preserved during that process. As a consequence it is feasible to reproduce the transfer function of propagation by means of a liquid crystal SLM which modulates the phase of the reflected light. With the Fourier transform properties of the lens and the focal length f we can substitute $\eta = u\,(\lambda f)^{-1}$ and $\zeta = v\,(\lambda f)^{-1}$ to arrive at the distribution to be generated by the SLM

$$h_z(u, v) = \exp\left[ikz\sqrt{1 - f^{-2}(u^2 + v^2)}\right], \tag{7.18}$$

where u and v are coordinates in the SLM domain. As an example for typical results obtained from phase retrieval, Fig. 7.5 shows a sequence of four intensity

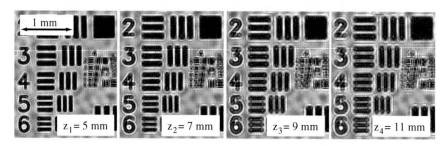

Fig. 7.5 Experimental results of phase retrieval obtained using the setup shown in Fig. 7.4 and an U.S. Air Force resolution test chart as object

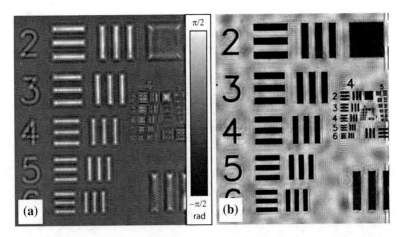

Fig. 7.6 **a** The phase distribution obtained from phase retrieval after 20 iterations of the extended Gerchberg-Saxton scheme and **b** the amplitude of the wavefield after numerical propagation to the focal plane

distributions captured using the above setup. The object was an U.S. Air Force resolution test chart (MIL-STD-150A standard) which has been investigated in transmission using LED light at $\lambda = 625$ nm and lenses with focal lengths of $f = 150$ mm. In the examples shown, the SLM generated four propagation transfer functions from $z_1 = 5$ mm to $z_4 = 11$ mm in steps of 2 mm. As a consequence the intensities appear to be blurred representations of the objects image, even though the object was placed in the front focal plane of the first lens.

In Fig. 7.6a the phase distribution across the plane at z_1 is shown after 20 iterations of the extended Gerchberg-Saxton scheme introduced above, i.e. numerically propagating through the planes while exchanging the amplitude by the measured one and preserving the phase function. In order to verify whether the obtained phase is correct, Fig. 7.6b shows the amplitude of the wavefield after numerically propagating it by 5 mm from z_1 to the focal plane using the *angular spectrum method*. Even very fine details are resolved proving that the phase of the wavefield has been determined accurately.

The setup can also be used for shape [266] and deformation [7] measurements. Here, the short acquisition time provided by the SLM has the great advantage that quasi static scenes can be investigated. This has made it feasible to apply phase retrieval to non-destructive testing with thermal load. The results are fully comparable with the displacement measurements obtained from digital holographic interferometry, as introduced in Sect. 4.1. Yet phase retrieval is much more tolerant against environmental disturbances, which is a major benefit with respect to industrial applications.

As an example we present the investigation of a *carbon fibre reinforced plastics* (CFRP) panel by means of thermal load. The panel, of which the front is seen in Fig. 7.7a, is part of an airplane fuselage. The task is to find the patch shown in

Fig. 7.7 Non-destructive testing of CFRP panels by means of phase retrieval: **a** *Front side* of the panel and **b** corresponding *reverse side* with a patch attached to it

Fig. 7.7b attached to the reverse side by looking at the front of the panel during thermal excitation. For the investigations an objective lens was used to produce images of the macroscopic object. The image plane of the objective lens was arranged to coincide with the object plane of the setup in Fig. 7.4 and a laser with wavelength $\lambda = 532$ nm was used to illuminate the specimen. While the object was in the relaxed state, a sequence of eight intensities were recorded, which correspond to propagation distances of $z_1 = 0$ mm to $z_8 = 1.4$ mm in steps of 0.2 mm. Note that the steps are comparably small because of the significant field curvature across the image plane of the lens objective.

The phase distribution of the relaxed state φ_1 has been obtained after 50 iterations. Subsequent to the first measurement, the object was thermally excited. The surface was heated by $\Delta T = 24$ Kelvin using infrared radiation. Another set of 8 intensities were recorded and the phase distribution of the loaded state φ_2 determined. As seen from Sect. 4.1, the fringes formed by the phase difference $\Delta \varphi = \varphi_1 - \varphi_2$ shown in Fig. 7.8a are proportional to the surface deformation due to the thermal expansion. They are dominated by a coarse deformation of the entire CFRP panel. After numerical compensation for this spatially low varying term, the foot print of the patch on the reverse side can be clearly identified from the inhomogeneous deformation of the front side material, as seen from Fig. 7.8b.

Finally, we would like to mention two prominent generalizations of the Gerchberg-Saxton approach which are called the *Error Reduction* method and the *Hybrid Input-Ouput* method (HIO) which both have been introduced by Fienup [61]. The idea behind these techniques is to apply the Fourier domain constraint in the same way like in the original Gerchberg-Saxton algorithm but to combine it with a different kind of a priori knowledge about the wavefield in the corresponding spatial domain. Important cases of pre-knowledge are non-negativity of an image, as found in astronomic speckle interferometry, or support limitations of $E_1(x,y)$, i.e. an aperture exhibiting a bounded area S outside of which the amplitude of the wavefield is known to drop to (or close to) zero. The generalized working principle of both methods is shown by the flow chart in Fig. 7.9.

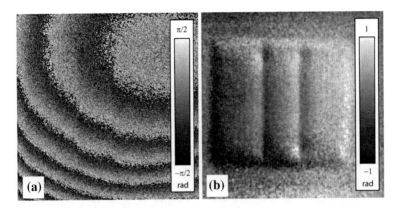

Fig. 7.8 Experimental results for non-destructive testing under thermal load by means of phase retrieval: **a** fringes indicating a strong deformation of a CFRP panel due to thermal load and **b** foot print of a patch on the *reverse side* of the panel identified by an inhomogeneous deformation on the *front side* [7]

Either of the algorithms can be interpreted as a non-linear system in which an input distribution $f_1(x,y)$ is processed by exchanging the amplitude in the Fourier domain by the square root of the measured intensity I_2, leading to an output distribution $f_1'(x, y)$. This operation is equivalent to the Fourier part of the Gerchberg-Saxton algorithm which can be described by the projection operator P_2 used in Eq. (7.13). Finally, the object constraints are applied based on the pre-knowledge.

Both methods work iteratively, starting with a random distribution $f_1^{(0)}(x,y)$ for example, but differ by the way the object constraints are applied. Let us denote the current iterate by $f_1^{(k)}(x, y)$. In the error reduction method, the next iterate $f_1^{(k+1)}(x, y)$ is computed from the projection $f_1'^{(k)}(x, y)$ by means of

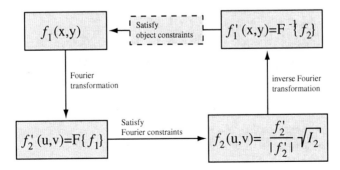

Fig. 7.9 Flow chart of the error reduction method and the hybrid input output method (*HIO*) introduced by Fienup. The index k usually indicating the number of iteration was left away for the sake of clarity here

$$f_1^{(k+1)}(x,y) = \begin{cases} f_1'^{(k)}(x,y), & (x,y) \notin \gamma \\ 0, & (x,y) \in \gamma \end{cases}. \qquad (7.19)$$

Here, γ is the set of points at which the projection $f_1'^{(k)}(x,y)$ violates the object constraints. It is straight forward to verify that for the above mentioned examples of non-negativity and/or limited support of the wavefield $E_1(x,y)$, Eq. (7.19) indeed is a projection in the strict sense of Eq. (7.12). Even further, both examples constitute a convex set in the sense of Eq. (7.11). In those cases the error reduction method can be described by 2 subsequent projections, similarly to Eq. (7.13). As a consequence, Eq. (7.15) holds and the method is guaranteed to reduce the summed distance error (or to stagnate), which is the name giving property. However, one of the main drawbacks is that after a few iterations the system is observed to only show very slow convergence rates.

In order to improve the convergence rate, the hybrid input output method was developed, according to which the next iterate $f_1^{(k+1)}(x,y)$ is calculated from the projection $f_1'^{(k)}(x,y)$ using

$$f_1^{(k+1)}(x,y) = \begin{cases} f_1'^{(k)}(x,y), & (x,y) \notin \gamma \\ f_1^{(k)}(x,y) - \beta \cdot f_1'^{(k)}(x,y), & (x,y) \in \gamma \end{cases}, \qquad (7.20)$$

where β is referred to as the feedback parameter. According to Fienup [60] a value between 0.5 and 1.0 works well in most cases. The hybrid input output method is observed to converge much faster to a reasonable solution than the error reduction algorithm and even escapes local minima in some cases. Having said that, it is difficult to conjecture a realistic case in which Eq. (7.20) constitutes a projection in the sense of Eq. (7.12) and we cannot guarantee convergence or even uniqueness of the solutions based on the formalism of generalized projections behind Eq. (7.15). However, we encourage the interested reader to have a look at the work of Bauschke et al. [12] who explained the success of the method in terms of classical convex optimization methods. Another intuitive explanation of the convergence properties is reported by Takajo et al. [234].

Recently, Fienup reported an update of the HIO method which he named continuous hybrid input output (CHIO), in order to overcome problems of the algorithm with oscillating solutions [60]. The problems arise because of the discontinuous behaviour of Eq. (7.20) in cases in which small changes of a particular value of $f_1'^{(k)}(x,y)$ decide whether it is in compliance with the constraints or not. A good example is the case in which the estimate $f_1^{(k)}(x,y)$ shall be a positive real valued function. To illustrate this case, Fig. 7.10 shows for a specific position (x_0, y_0) the value of the next iterate $f_1^{(k+1)}$ in dependence of the projection $f_1'^{(k+1)}$ and the current iterate $f_1^{(k)}$ for both, the error reduction method and the HIO method as obtained from Eqs. (7.19) and (7.20) respectively.

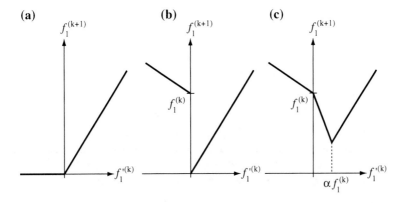

Fig. 7.10 The value of the next iterate $f_1^{(k+1)}$ depending on the value of the projection $f_1'^{(k)}$ subject to the non-negativity constraint for **a** the error reduction method, **b** the standard HIO method and **c** the continuous HIO method [60]

In case of the error reduction method the successor $f_1^{(k+1)}$ simply becomes zero when the projection $f_1'^{(k)}$ yields a negative value as seen from the diagram in Fig. 7.10a. In the same situation, HIO has a discontinuous step, hence an infinitesimal variation of $f_1'^{(k)}$ in this region can drastically affect the resulting value of $f_1^{(k+1)}$. This can be seen from Fig. 7.10b. The aim of CHIO is to provide a continuous transition, which in case of non-negativity can be realized by the following modification of Eq. (7.20):

$$f_1^{(k+1)}(x,y) = \begin{cases} f_1'^{(k)}(x,y), & \alpha f_1^{(k)}(x,y) \le f_1'^{(k)}(x,y) \\ f_1^{(k)}(x,y) - \frac{1-\alpha}{\alpha} \cdot f_1'^{(k)}(x,y), & 0 \le f_1'^{(k)}(x,y) \le \alpha f_1^{(k)}(x,y) , \\ f_1^{(k)}(x,y) - \beta \cdot f_1'^{(k)}(x,y), & \text{otherwise} \end{cases} \quad (7.21)$$

where α is a second feedback parameter. The behaviour of CHIO with respect to the above example is shown is Fig. 7.10c. The algorithm causes fewer problems with oscillations and converges even faster when compared to HIO.

There also exist a few variants based on the above projection based approaches, of which a detailed discussion is beyond the scope of this introduction. For the interested reader we add some reference to the *hybrid projection-reflection* algorithm (HPR) which is a special case of CHIO with $\alpha = 1/(1 + \beta)$ and has been design specifically for non-negativity constraints [13], the *difference map* approach (DM) [50] which is an attempt to unify some projection based algorithms, the averaged successive reflections algorithm (ASR) [12] and the *relaxed averaged alternating reflectors* algorithm (RAAR) [149] which again applies the theory of convex optimization to the non-convex problem of phase retrieval in order to make it mathematically tractable.

7.2.2 Gradient Search Methods

An alternative way for minimization of the functional in Eq. (7.3) is to employ gradient search methods [217]. In the following we would like to detail this concept by means of the *steepest descent gradient* method, since it is the most instructive and intuitive gradient technique.

Similarly to the Gerchberg-Saxton type of approach, the basic idea is to start with an arbitrary initial guess for the phase distribution in the first plane $\varphi_1^{(0)}$ and to iteratively refine the estimate. However, in contrast to projection based algorithms the next iteration is calculated by directly following the negative gradient of the objective function L with respect to the function values of the current phase estimate $\varphi_1^{(k)}(x,y)$:

$$\varphi_1^{(k+1)}(x,y) = \varphi_1^{(k)}(x,y) - \alpha^{(k)} \cdot \nabla L^{(k)}(x,y). \tag{7.22}$$

Here $\nabla L(x,y)$ denotes the partial derivative of the functional L after the values of the phase estimate $\varphi_1^{(k)}$ at position (x,y)

$$\nabla L^{(k)}(x,y) = \frac{\partial L}{\partial \varphi_1^{(k)}(x,y)}. \tag{7.23}$$

Since the gradient always points towards the greatest increase, following its opposite direction will reduce the value of L. The scalar parameter $\alpha^{(k)}$ is called the step size, which is the distance in Hilbert space that we are willing to follow the negative gradient. It is often optimized numerically by a trial and error method. The most primitive algorithms simply calculate the values of $L(\varphi_1^{(k+1)})$ for different values of $\alpha^{(k)}$ and finally decide for that value of $\alpha^{(k)}$ which yields the smallest value for $L(\varphi_1^{(k+1)})$.

Please, note that because $f_1^{(k)}$ and $\varphi_1^{(k)}$ are two dimensional functions, the gradient is also a two dimensional function and may be complicated to determine. For numerically efficient implementation it is therefore convenient to have an explicit expression for the gradient. Inserting Eq. (7.3) into Eq. (7.23) yields [93]

$$\nabla L^{(k)} = \text{Im}\left[f_1^{(k)} \cdot \sum_n \psi_n^{(k)*} \right], \tag{7.24}$$

where

$$\psi_n^{(k)} = P^{-1}\left\{ \sqrt{I_n} \frac{f_n^{(k)}}{\left| f_n^{(k)} \right|} ; z_n - z_1 \right\}. \tag{7.25}$$

Here, P^{-1} is the inverse of the propagation operator and $f_1^{(k)} = A_1 \exp(i\varphi_1^{(k)})$ is the estimate of the complex amplitude in the first plane. Let us discuss the structure of Eq. (7.25) a bit further. The $f_n^{(k)}$ are the estimates of the wavefield in the nth plane based on the kth iteration $f_1^{(k)}$. They can be expressed by means of the propagation operator

$$f_n^{(k)} = P\left\{f_1^{(k)}; z_n - z_1\right\}. \tag{7.26}$$

According to Eq. (7.25) the amplitude of the propagated wavefield $f_n^{(k)}$ is exchanged by the square root of the measured intensity and then propagated back again to the first plane by means of the inverse of the propagation operator. Hence, the function $\psi_n^{(k)}$ describes the projection of the current estimate $f_1^{(k)}$ onto the set of functions having the intensity I_n in the nth plane. If we generalize the projection operator in analogy to Eq. (7.13), we can write:

$$\psi_n^{(k)} = P_n\left\{f_1^{(k)}\right\}. \tag{7.27}$$

This remarkable result indicates a close relation between projection based and gradient based methods. It also shows that the gradient can be calculated most efficiently by simply adding a series of forth and back propagations while exchanging the amplitudes by the ones obtained from the measured intensities.

Alternatively, gradient based methods can be employed to optimize for the complex values of $f_1^{(k)}$ rather than for the phase values. In this case, the gradient is defined by

$$\nabla L^{(k)}(x, y) = \frac{\partial L}{\partial f_1^{(k)}(x, y)}. \tag{7.28}$$

The main advantage of this approach is that the amplitudes in the first plane are varied as well, which is a benefit over simply accepting the square root of the noisy intensity observations. Inserting Eq. (7.3) into Eq. (7.28) yields the analytic form of the gradient [60]

$$\nabla L^{(k)} = -2\sum_n \chi_n^{(k)}, \tag{7.29}$$

where

$$\chi_n^{(k)} = P^{-1}\left\{\sqrt{I_n}\,\frac{f_n^{(k)}}{\left|f_n^{(k)}\right|} - f_n^{(k)}; z_n - z_1\right\}. \tag{7.30}$$

In analogy to Eq. (7.22), the gradient can be used to calculate the successor of the complex amplitude by

$$f_1^{(k+1)}(x,y) = f_1^{(k)}(x,y) - \alpha^{(k)} \cdot \nabla L^{(k)}(x,y). \tag{7.31}$$

Finally, we should mention that the same gradients Eqs. (7.24) and (7.29) can be used to calculate the search directions (conjugate gradients) for the conjugate gradient (CG) method, which converges in many cases much faster than the simple steepest descent approach outlined above.

7.2.3 Deterministic Methods

Deterministic methods of phase retrieval aim at calculating the phase directly from a set of intensities rather than iteratively approximating a solution which minimizes Eq. (7.3). The most prominent technique in this field is based on the so called transport-of-intensity equation (TIE) introduced by Teague [238] which provides an analytic relation between phase and intensity in monochromatic light. The TIE is based on the Fresnel approximation and its application for phase retrieval is limited to cases in which the intensity across the observation plane is constant. In optical metrology this restricts any method using the TIE to pure phase objects being illuminated with homogenous light, e.g. transparent specimen illuminated by a plane wave. As a consequence all wavefields exhibiting vortices or diffraction effects cannot be investigated by this method. However, the big advantage of TIE based techniques is that they are computationally very efficient and in principle only require the acquisition of the intensity across two parallel planes separated along the main axis of propagation. In the following, we will derive the TIE from the Helmholtz-Equation, discuss its properties and finally show an example of application.

The main assumption of the TIE is the paraxial, or Fresnel approximation, i.e. the light rays are assumed to travel along or close to the main axis of propagation. Without loss of generality we may choose the positive z-axis to be the main direction of propagation. In this case we can express the complex amplitude of a monochromatic wavefield at any point $r = (x,y,z)$ in space by means of

$$u(r) = f(r) \cdot \exp(ikz), \tag{7.32}$$

where $f(r)$ is a complex valued function which is slowly varying along the z-axis. We can insert Eq. (7.32) into the Helmholtz-Equation

$$\left(\nabla^2 + k^2\right)u(r) = 0, \tag{7.33}$$

to verify what requirement the complex function $f(r)$ has to fulfil in order to satisfy it. With the assumption that $f(r)$ is slowly varying along the z-direction we may set its second derivative after z to zero and yield

$$\left(\frac{1}{2k}\nabla_T^2 + i\frac{\partial}{\partial z}\right)f(r) = 0, \tag{7.34}$$

where we introduced the transverse part of the Laplace-Operator ∇_T in the (x,y)-plane

$$\nabla_T^2 = \frac{\partial^2}{\partial x^2} + \frac{\partial^2}{\partial y^2}. \tag{7.35}$$

The differential equation Eq. (7.34) represents the paraxial approximation of the Helmholtz-Equation. In many cases, it is more convenient to express it in terms of $u(r)$ rather than $f(r)$ by means of the parabolic equation

$$\left(\frac{1}{2k}\nabla_T^2 + i\frac{\partial}{\partial z} + k\right)u(r) = \psi\{u(r)\} = 0. \tag{7.36}$$

which can be easily verified by inserting Eq. (7.32) and using Eq. (7.34). Here, we introduced the operator ψ for the sake of brevity. Teague made the following statement using the above parabolic equation

$$u^*(r) \cdot \psi\{u(r)\} - u(r) \cdot \psi\{u^*(r)\} = 0. \tag{7.37}$$

By Inserting Eq. (7.32) and rearranging the terms it is straight forward to yield the transport-of-intensity equation

$$k\frac{\partial}{\partial z}I(r) = -\nabla_T \cdot I(r)\nabla_T\varphi(r). \tag{7.38}$$

The TIE is a differential equation that relates the phase $\varphi(r)$ of the wavefield to its intensity $I(r)$. In case that the intensity is known to be different from zero across the entire observation plane it can be solved numerically for example by means of orthogonal series expansion [77] or multi-grid methods [8], which is computationally demanding. A much more efficient solution is achieved when $I(x,y,z_0) = I_0$ can be assumed, i.e. the intensity in the observation plane at position z_0 is constant. In this case we yield the following Poisson equation:

$$-\frac{k}{I_0}\frac{\partial}{\partial z}I(r) = \nabla_T^2\varphi(r). \tag{7.39}$$

However, even though $I(\mathbf{r})$ is constant across the (x,y)-plane at z_0, it still can vary in the z-direction which means that its derivative after z will in general be different from zero. To solve the Poisson equation in a numerically efficient way we make use of the fact that, according to Eq. (A.13) in Appendix A2, derivation of a function can be described by means of a transfer function in the frequency domain. This can be exemplified by considering the Fourier transform $G(v)$ of an arbitrary complex function $g(x)$

$$g(x) = \int_{-\infty}^{\infty} G(v)\exp(i2\pi xv)dv. \tag{7.40}$$

Derivation of g after x yields

$$\frac{\partial g}{\partial x} = \frac{\partial}{\partial x}\left(\int_{-\infty}^{\infty} G(v)\exp(i2\pi xv)dv\right) = \int_{-\infty}^{\infty} G(v)H(v)\exp(i2\pi xv)dv, \tag{7.41}$$

with the linear term $H(v) = i2\pi v$ representing the transfer function of differentiation. By analogy with the application of a simple derivative in Eq. (7.41) we can also apply the transverse Laplacian Eq. (7.35) to Eq. (7.40) and yield the corresponding transfer function

$$T(v, \xi) = -4\pi^2\left(v^2 + \xi^2\right). \tag{7.42}$$

This implies that the Poisson Equation (7.39) can be solved by simply applying the inverse filter $T^{-1}(v,\xi)$ in the Fourier domain. However, T^{-1} has a pole at $v = \xi = 0$ and if noisy measurements are at hand it is necessary to find an optimum filter in a least-squares sense. We will define an objective function based on the Tikhonov regularization which constraints the solution to the one with minimum norm:

$$L(f) = \left\|\nabla_T^2 f(x, y) - m_\varepsilon(x, y)\right\|^2 + \alpha\|f(x, y)\|^2. \tag{7.43}$$

Here, $m_\varepsilon(x,y) = m(x,y) + \varepsilon(x,y)$ represents the measurement, where $m(x,y)$ stands for the left hand side of Eq. (7.38) and $\varepsilon(x,y)$ is additive noise. The regularization parameter α has to be chosen inversely proportional to the signal to noise ratio. The least-squares solution $\varphi_{LSE}(x,y)$ is provided by the minimum of the functional L with respect to f

$$\varphi_{LSE}(x, y) = \min_f L. \tag{7.44}$$

Calculation of the gradient of L after the function values of f is non-trivial because of the transverse Laplacian. We therefore employ Parseval's theorem and perform the optimization in the Fourier domain

$$L(F) = \|F(v, \xi)T(v, \xi) - M_\varepsilon(v, \xi)\|^2 + \alpha\|F(v, \xi)\|^2, \qquad (7.45)$$

where capitalization denotes the Fourier transform of their lower case counterparts. To find the optimum F which minimizes L we set the gradient to zero:

$$\frac{\partial L}{\partial F} = \frac{\partial L}{\partial F_R} + \mathrm{i}\frac{\partial L}{\partial F_I} = 2T^*(F \cdot T - M_\varepsilon) + 2\alpha F = 0, \qquad (7.46)$$

where $F_R(v,\xi)$ and $F_I(v,\xi)$ are the real and the imaginary part of $F(v,\xi)$. Finally, by rearranging for F and using the fact that according to Eq. (7.42) $T^* = T$, we obtain the optimum filter T_{INV} which can be applied in the Fourier domain in order to recover the Fourier transform of the phase in a least squares sense:

$$F(v, \xi) = \frac{T^*(v, \xi)}{|T(v, \xi)|^2 + \alpha} M_\varepsilon(v, \xi) = T_{INV}(v, \xi; \alpha) \cdot M_\varepsilon(v, \xi). \qquad (7.47)$$

Hence, we can determine the least-squares solution for the phase by means of two Fourier transforms and multiplication of the inverse filter

$$\varphi_{LSE}(x, y) = -\frac{k}{I_0}\mathcal{F}^{-1}\left\{\mathcal{F}\left\{\frac{\partial I(x, y, z_0)}{\partial z}\right\} \cdot \frac{T^*(v, \xi)}{|T(v, \xi)|^2 + \alpha}\right\}. \qquad (7.48)$$

The values of this solution are not wrapped in the interval $[-\pi, \pi]$. Instead, φ_{LSE} constitutes a smooth function which can be related to a wavefront. Furthermore, due to the use of the Fourier transform, the solution is inherently assumed to be periodic in all cases in which the values of m_ε are merely known across a limited domain.

A crucial step in the application of TIE based techniques is the determination of the derivative of the intensity along the z-axis, which serves as the input data to the method. It cannot be measured directly and has to be derived from a set of intensity observations. The most straight forward approach is to make a finite approximation employing a difference quotient based on two measurements at positions z_+ and z_-:

$$\frac{\partial I(x, y, z_0)}{\partial z} \approx \frac{I(x, y, z_+) - I(x, y, z_-)}{z_{+1} - z_-}. \qquad (7.49)$$

If any of the positions z_+ or z_- is selected to equal z_0, i.e. the position of the observation plane, only two measurements are required. However, the approximation is better if the measurements are symmetrically distributed around the observation plane, requiring at least three intensity measurements at z_-, z_0 and z_+ for the phase reconstruction in Eq. (7.48). The choice of the distance $\Delta z = z_+ - z_-$ is a trade-off. Small distances will yield better approximation of the difference quotient, while larger distances will improve the detection of intensity changes. In practice, a distance in the range of the diffraction limited depth of focus of the imaging system has proven useful [225]. Even better approximations can be achieved for example

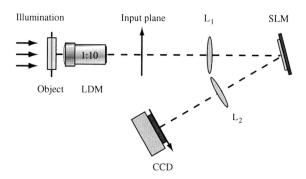

Fig. 7.11 Measuring the relative optical path by means of quantitative phase contrast microscopy using a TIE based technique. The setup is a combination between a long distance microscope objective (*LDM*) and the SLM based phase retrieval setup introduced before. The object is a gradient index fibre in a *box-shaped* container filled with an index matching liquid. The object is illuminated by collimated light coming from a fibre coupled LED with central wavelength $\lambda = 625$ nm

by polynomial fitting [247] when intensity measurements in multiple planes are considered.

In the following we will present an example of application from the field of quantitative phase contrast microscopy. The setup is shown in Fig. 7.11. It is a combination of a long distance microscope objective with a magnification of $M = 1:10$ and the 4f-configuration introduced in Fig. 7.4. The image plane of the microscope objective coincides with the input plane of the 4f-setup. The object is a gradient index (GRIN) fibre with a core diameter of $D = 62.5$ µm. The fibre is inserted into a box-shaped container filled with a liquid that matches the refractive index of the cladding. The aim of the experiment is to measure the relative optical path of light transmitting the container. It is expected that the fibre core will delay the light significantly. From the results it is possible to reason back on the properties of the fibre core, such as symmetry, homogeneity along the fibre axis or, if a cylindric shape can be assumed, even the refractive index distribution. The light source is a fibre coupled LED which emits light at $\lambda = 625$ nm.

In Fig. 7.12 we see the three intensity measurements at positions $z_- = -10$ µm, $z_0 = 0$ µm and $z_+ = 10$ µm relative to the focal plane. The microscope objective has a numerical aperture of $N_A = 0.28$, so that the distance $\Delta z = z_+ - z_- = 20$ µm is well beyond the diffraction limited depth of focus

$$l_R = \frac{\lambda}{2N_A^2} \qquad (7.50)$$

of $l_R = 3.9$ µm. The method is quite robust against dirt and dust particles on the optics which are comparably far away from the focal plane, because their appearance does not change significantly due to the defocusing procedure and only differences of the captured intensities are considered for the numerical integration. In

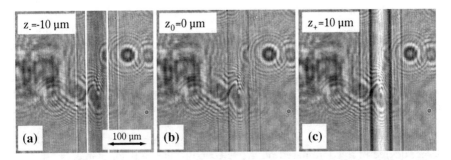

Fig. 7.12 A sequence of microscope images of the fibre in the container with relative distance of a $z_- = -10$ μm, **b** $z_0 = 0$ μm and **c** $z_+ = 10$ μm to the focal plane of the microscope objective

Fig. 7.13a the result for the relative optical path as obtained from Eq. (7.48) is shown, where the difference quotient Eq. (7.49) has been used to approximate the derivative of the intensity after z, and the average intensity in plane z_0 has been used as I_0. The influence of the fibre core on the optical path is clearly seen and even the cladding is visible. In the background we see low frequent deviations which are assumed to have two reasons. First of all, the frequency dependent signal to noise ratio is proportional to the transfer function $T(v,\xi)$ which is zero at the dc-term and very small for low frequencies, as seen from Eq. (7.42).

However, a careful analysis shows that the dominant term of the low frequent background is of second order and that the maximum is close to the center of the observation plane. This indicates that the strong image field curvature of the microscope objective has caused the deviation. However, apart from this artefact the field curvature is not detected at all because it has no significant effect on the

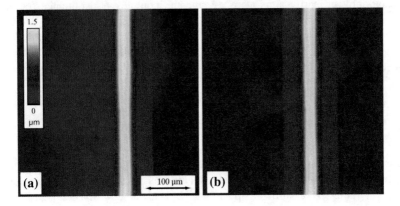

Fig. 7.13 Relative optical path of light transmitting a fibre measured by means of a TIE based technique: **a** result of the numerical integration Eq. (7.48) and **b** after compensation of the low frequent background by subtraction of a 5th order polynomial fit. The fit coefficients have been calculated from the regions *left* and *right* to the fibre which are known to be constant

intensity, revealing the weakness of the TIE approach with respect to smooth wavefronts. In Fig. 7.13b we removed the background variations by subtraction of a polynomial fit. The fit coefficients have been calculated from the regions left and right to the fibre which were assumed constant.

Finally, for the interested reader we would like to refer to some interesting alternative approaches to tackle the phase retrieval problem. Frank et al. [62], suggest using the first Green's identity and the Green's function to calculate the phase. The method requires the same observations and has almost the same properties like techniques using the TIE. The main advantage is that it offers a larger degree of freedom based on the selection of the particular Green's function. This can be used to adapt the phase retrieval problem to a priori knowledge about the behavior of the phase along the edges of the observation area for example.

The work of Kolenovic [120] sheds light into the relationship between intensity of a wavefield and the full vectorial representation of the phase gradient rather than only the absolute value of the gradient as provided by the TIE. From his results it is possible to derive the important statement that in the general case of an arbitrary wavefield at least four intensity measurements are required to yield a unique solution.

Agour introduced a hybrid approach by combining the advantages of deterministic and iterative methods [267]. It is based on deriving a set of criteria from the Helmholtz equation that are used to invoke additional constrains. The technique is much faster converging compared to standard iterative phase retrieval and also finds reasonable solutions if only a minimum of 4 intensity observations are at hand.

Recently, the scheme of *Ptychography* has been reported [192], which shares large methodological similarities with phase retrieval. It is an iterative approach based on intensity measurements of light scattered by a specimen. The intensity observations correspond to different transverse positions of the sample, i.e. the sample is laterally moved between the recordings. In contrast to phase retrieval, Ptychography primarily aims at recovering the complex transmittance (or object function) of the sample rather than the complex amplitude of a wavefield. However, given the characteristics of the illumination, the complex transmittance of the object can be directly related to the complex amplitude of the wavefield directly behind the object.

7.3 Shear Interferometry for Wavefield Sensing

Shear interferometry is an interferometric method, but in contrast to classical interferometry the wavefield under investigation is not superposed by a reference wave but rather by a shifted copy of itself. The name *shear* denotes the shift. Consequently, the interference pattern observed in the sensor plane of a shear interferometer is given by

$$I(\mathbf{r}) = |u(\mathbf{r})|^2 + |u(\mathbf{r}+\mathbf{s})|^2 + 2\mathrm{Re}\{u^*(\mathbf{r}) \cdot u(\mathbf{r}+\mathbf{s})\}, \qquad (7.51)$$

where $u(\mathbf{r})$ is the complex amplitude of the investigated wavefield, $\mathbf{s}$ is the shear vector and $\mathbf{r}$ is a vector in the sensor domain. Many setups allow for introduction of an artificial phase step between the wavefields, so that any phase shifting technique can be used to extract the product in the third term of Eq. (7.51)

$$M(\mathbf{r}) = u^*(\mathbf{r})u(\mathbf{r}+\mathbf{s}) = a(\mathbf{r})a(\mathbf{r}+\mathbf{s}) \cdot \exp[i\Delta\varphi(\mathbf{r})], \qquad (7.52)$$

with $\Delta\varphi(\mathbf{r}) = \varphi(\mathbf{r}+\mathbf{s}) - \varphi(\mathbf{r})$ being the difference of the phase values at two points separated by the shear. The great benefit of shear interferometry over other interferometric techniques is that it is very robust against environmental disturbances, such as mechanical vibrations and thermal fluctuations. Due to the common path principle it also has remarkably low demands with respect to the temporal and spatial coherence of the investigated light. Indeed, the only requirement is that the mutual intensity $G(\mathbf{r},\mathbf{r}+\mathbf{s})$ in the sensor domain at positions separated by the shear is significantly different from zero

$$G(\mathbf{r},\mathbf{r}+\mathbf{s}) = \langle u^*(\mathbf{r}) \cdot u(\mathbf{r}+\mathbf{s})\rangle_T > 0. \qquad (7.53)$$

Here, $\langle ...\rangle_T$ denotes the time average. On the other hand, the great disadvantage of shear interferometry is that the measurement only provides compound information about the complex amplitudes at points separated by the shear. If the goal of the investigation is associated with identification of the underlying wavefield itself, this requires sophisticated numerical post processing of the measured data in order to recover parts of the wavefield or even the entire complex amplitude.

Following our approach in phase retrieval, we will formulate the corresponding inverse problems by means of objective functions which have to be minimized in a least-squares sense. If the goal is solely recovering the phase distribution of the wavefield, the task is to minimize

$$L(f) = \sum_n \|w_n(\mathbf{r}) \cdot ([f(\mathbf{r}+\mathbf{s}_n) - f(\mathbf{r})] - \Delta\varphi_n(\mathbf{r}))\|^2. \qquad (7.54)$$

The subscript n denotes the number of the measurement where different shears $\mathbf{s}_n$ are evaluated in combination. The weighting function $w_n(\mathbf{r})$ can be used to express confidence in the measured phase differences $\Delta\varphi_n$ or to simply mark invalid regions. The objective function Eq. (7.54) is well suited to cases in which the phase is known to constitute a smooth and continuous wavefront, i.e. $f(\mathbf{r})$ is a potential function. If the measured differences exceed the range $[-\pi,\pi]$, they have to be unwrapped prior to the inversion process. The main applications of this scheme are shape measurement for optical components and wavefront analysis for adaptive optics.

If the entire complex amplitude is to be recovered we choose the following objective function:

$$L(f) = \sum_n \|M_n(r) - f^*(r)f(r + s_n)\|^2. \tag{7.55}$$

Here the function $f(r)$ is complex valued. Techniques minimizing Eq. (7.55) are even capable of recovering wavefields with so-called *phase singularities*. Phase singularities appear at all positions where the amplitude of a wavefield equals zero, such as in speckle fields or diffracted light in general. In the vicinity of a phase singularity, the phase distribution has a vortex-like structure and takes all values from $-\pi$ to π [11]. Consequently, the corresponding wavefields do not have smooth wavefronts. The complex amplitudes obtained from minimization of Eq. (7.55) are comparable to the results of phase shifting digital holography and allows for numerical refocusing for example. In the following chapters we will explore both problems Eqs. (7.54) and (7.55) separately and give examples of applications.

7.3.1 Wavefront Reconstruction

In wavefront reconstruction [88], it is assumed that the phase distribution $\varphi(r)$ to be recovered is a smooth and continuous function. This is a very strong constraint providing great benefits in the inversion process. Depending on the shape of the pupil function and the weighting we may distinguish two different approaches. If the wavefront is to be recovered across a rectangular pupil and, if weighting through $w_n(r)$ is not considered, the above mentioned smoothness constraint gives rise to computationally very efficient direct Fourier methods to find an optimum estimate of the wavefront [65]. If, however, the pupil function is irregularly shaped or if weighting through $w_n(r)$ is required in general, Eq. (7.54) can still be iteratively solved by means of gradient based methods.

Before we start discussing both of these options in detail, let us first consider the uniqueness of the solution. The measurement described by Eq. (7.52) only provides differences of the phase between two positions separated by the shear. Hence, if we for example chose a periodic phase distribution $\varphi_p(r)$ which has a period that equals the shear, i.e. $\varphi_p(r + s) = \varphi_p(r)$ we find that everywhere

$$\Delta\varphi(r) = 0 \tag{7.56}$$

Apparently, shear periodic parts of the phase distribution are lost during the shearing process. This implies that inversion based on measurements with a single shear is not unique. In the past few decades, this has led to the development of inversion methods for the recovery of phase distributions that consider strong assumptions, such as a minimum curvature of the underlying wavefront [211] or a priori knowledge about its specific structure [19, 54].

In the general case, we have to evaluate measurements with varying direction and magnitude of the shear. In order to derive rules for the selection of the shears it

Fig. 7.14 The absolute value of the shear transfer function $|H(v)|$. The roots correspond to frequencies which are not transferred by the shear process

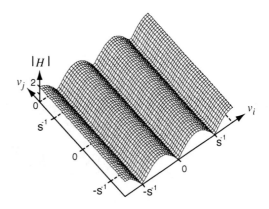

is instructive to visualize the problem of uniqueness in Fourier space. For this, we will reformulate the phase difference $\Delta\varphi$ as a convolution with two Dirac distributions:

$$\Delta\varphi(\mathbf{r}) = \varphi(\mathbf{r}+\mathbf{s}) - \varphi(\mathbf{r}) = \varphi(\mathbf{r}) \otimes [\delta(\mathbf{r}+\mathbf{s}) - \delta(\mathbf{r})] \qquad (7.57)$$

Applying the convolution theorem [see Appendix A2, Eq. (A9)] we find for the Fourier transform

$$\mathcal{F}\{\Delta\varphi\} = \mathcal{F}\{\varphi\} \cdot \mathrm{i}2\sin(\pi\mathbf{s}\cdot\mathbf{v})\exp(\mathrm{i}\pi\mathbf{s}\cdot\mathbf{v}) = \mathcal{F}\{\varphi\} \cdot H(\mathbf{v}). \qquad (7.58)$$

Hence, in case of wavefront analysis, the shear process can be described as a linear shift invariant system with transfer function $H(\mathbf{v})$, where $\mathbf{v} = (v_i, v_j)^{\mathrm{T}}$ is a vector in the Fourier domain. The structure of the transfer function can be seen from Fig. 7.14 where the modulus $|H(\mathbf{v})|$ is depicted. The frequencies which are not transferred by the shear process correspond to the roots of $H(\mathbf{v})$. They constitute lines in the Fourier domain following

$$\mathbf{s}\cdot\mathbf{v} = n, \qquad (7.59)$$

where n is an integer number. Hence, the shear operation does not transfer any frequencies of the phase distribution that exhibit an integer number of full periods along the shear. For such frequencies, the result of the subtraction in Eq. (7.56) will always yield zero, regardless of the amplitude. Additionally, frequencies close to the roots in the Fourier domain will only be transferred with a low signal-to-noise ratio (SNR), which will have an effect on the smoothness of the inversion. It is worth noting that for frequencies exhibiting an integer number of full periods along the shear plus a half period, i.e.

$$\mathbf{s}\cdot\mathbf{v} = n + \frac{1}{2}, \qquad (7.60)$$

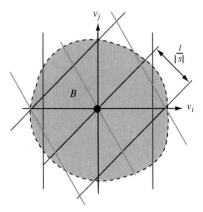

Fig. 7.15 Example for the selection of the shears: the diagram shows the assumed band B of a wavefront to be recovered. It is superposed with *lines*, where each *color* refers to the roots of one of three shear transfer functions. A mandatory requirement for the uniqueness of the inversion is that all three colors must not intersect at any point on the band

the transfer function has a value of 2. This means that the shear process is very sensitive to those frequencies.

Now we can choose the number, magnitude and the orientation of the shears based on the signal-to-noise ratio in Fourier space, where the primary goal is to avoid common roots of the corresponding transfer functions. The case corresponding to three shears is illustrated in Fig. 7.15. The diagram shows the assumed band B of the phase distribution φ to be recovered, i.e. those spatial frequencies which are known to have a modulus significantly different from zero. The coloured lines represent roots of transfer functions, where each colour is associated with one transfer function. To ensure uniqueness, it is necessary that at no point in the band do all of the colours intersect at the same time. An unavoidable exception to this rule is the dc-term in the origin. The dc-term will never be transferred by any shear operation. This has to be considered by the reconstruction process and simply refers to the well-accepted fact that any differential approach can only recover a signal up to an unknown constant. The same diagram can be used to ensure a sufficient signal to noise ratio for all of the frequencies. In this case, the shears have to be selected in a way that the intersections of the lines have maximum distance from each other. In practice, it has proven useful to select shears in orthogonal directions and relative prime magnitudes to each other.

If uniqueness up to a constant is ensured, we can recover the wavefront most elegantly across a rectangular pupil function by means of an inverse filter in the Fourier domain. However, we have to account for the expected pole at the dc-term. Therefore we slightly reformulate the objective function Eq. (7.54) and, similar to that in Sect. 7.2.3, add a Tikhonov-Regularization

$$L(f) = \sum_n \left\| [f(\mathbf{r} + \mathbf{s}_n) - f(\mathbf{r})] - m_{n,\varepsilon}(\mathbf{r}) \right\|^2 + \alpha \|f(\mathbf{r})\|^2 \qquad (7.61)$$

in order to select the solution with minimum norm. Here, $m_{n,\varepsilon}(\mathbf{r}) = \Delta\varphi_n(\mathbf{r}) + \varepsilon_n(\mathbf{r})$ denote the observations with additive noise $\varepsilon_n(\mathbf{r})$ and we assume no weighting, i.e. $w_n(\mathbf{r}) = 1$ everywhere. Minimization of Eq. (7.61) is straight forward in the Fourier domain. We therefore use Eq. (7.57) on $f(\mathbf{r})$ and employ Parseval's theorem

$$L(F) = \sum_n \left\| F(\mathbf{v})H_n(\mathbf{v}) - M_{n,\varepsilon}(\mathbf{v}) \right\|^2 + \alpha \|F(\mathbf{v})\|^2. \qquad (7.62)$$

To find the optimum F which minimizes L we calculate the gradient and set it to zero:

$$\frac{\partial L}{\partial F(\mathbf{v})} = 2 \sum_n H_n^* \left(F \cdot H_n - M_{n,\varepsilon} \right) + \alpha F = 0. \qquad (7.63)$$

Finally, solving the equation for $F(\mathbf{v})$ yields an analytic expression for the Fourier transform of the optimum wavefront in a least squares sense:

$$F(\mathbf{v}) = \frac{1}{\sum_n |H_n(\mathbf{v})|^2 + \alpha} \cdot \sum_n M_{n,\varepsilon}(\mathbf{v}) \cdot H_n^*(\mathbf{v}). \qquad (7.64)$$

From this result, the phase can be recovered by a simple inverse Fourier transform. The regularisation parameter α avoids over-fitting in close vicinity to the pole at $\mathbf{v} = 0$ and has to be chosen inversely proportional to the signal-to-noise ratio.

An implicit assumption made when using the Fourier transform in Eq. (7.64) is that the measured phase differences $\Delta\varphi(\mathbf{r})$ are known across the entire spatial domain. In practice, discrete sensing devices such as CCD cameras with a limited spatial extent are used, which means that the sheared representations of the wavefront are only known across a limited pupil function. In those cases we have to use the discrete Fourier transform (DFT) which inherently extends the measurements to periodic distributions, which we will denote by $\Delta\varphi_p(\mathbf{r})$. This causes serious problems if the area where $\Delta\varphi(\mathbf{r})$ is significantly different from zero is not spatially limited to the sensor device. In this case, the periodic function $\Delta\varphi_p(\mathbf{r})$ assumed by the DFT does not appear to be the result of a shearing process, because the phase differences across the boundary of the pupil function contain wavefront data from outside of it.

An elegant method to solve this problem is called *natural extension* and has been reported by Elster and Weingärtner [51]. It works on rectangular pupils with all shears aligned parallel to any of the edges. Additionally, the spatial extend of the rectangular pupil has to equal a multiple N of the shear-magnitude in the corresponding direction. The idea is to modify the phase values across the border region of $\Delta\varphi(\mathbf{r})$, letting the periodic functions $\Delta\varphi_p(\mathbf{r})$ appear to be sheared representations of a periodic wavefront $\varphi_p(r)$ that equals the wavefront $\varphi(r)$ across the pupil. In order to find the correct values for the modification, consider the fortunate situation in which the investigated wavefront $\varphi(r)$ is already inherently periodic, i.e. across the pupil area $\varphi_p(r) = \varphi(r)$ holds. An interesting property of these distributions is given by the sum along a set of points separated by the shear. It is straightforward to show that it yields zero, i.e.

$$\sum_{n=0}^{N-1} \varphi_P(r + n \cdot s) = 0. \tag{7.65}$$

In general, the wavefront under investigation is not inherently periodic, and the measured distributions are not expected to fulfill the constraint given by Eq. (7.65). However, in order to force periodicity, the coefficients located on the boundary of the pupil can be modified accordingly to arrive at the naturally extended distribution. In terms of the noisy observations $m_\varepsilon(r)$ we may write

$$m_P(r) = \begin{cases} -\sum_{n=0}^{N-2} m_\varepsilon(r + n \cdot s); & N|s| > \frac{s}{|s|} \cdot r \geq (N-1)|s| \\ m_\varepsilon(r); & \text{everywhere else} \end{cases}. \tag{7.66}$$

The results of Eq. (7.66) can be inserted into Eq. (7.64) in order to reconstruct the wavefront across the limited area.

In the following we will present results of a numerical simulation in order to demonstrate the method. In Fig. 7.16a, b we see amplitude and phase of a wavefield. The amplitude is a normalized Gaussian distribution which drops off to 4 % of its maximum value in the edges. The corresponding phase distribution has a checkerboard like appearance with a phase difference of $\pi/2$ between dark and

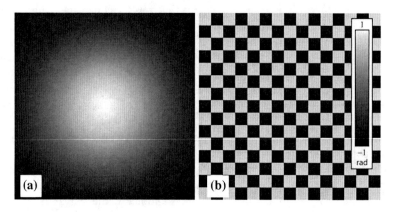

Fig. 7.16 Numerical simulation: **a** amplitude and **b** phase of the wavefield used for the numerical simulation of the shear process. The amplitude is a normalized Gaussian distribution which drops off to 4 % of its peak value in the borders. The phase has a checkerboard structure with steps of $\pi/2$ between the *dark* and *light fields*

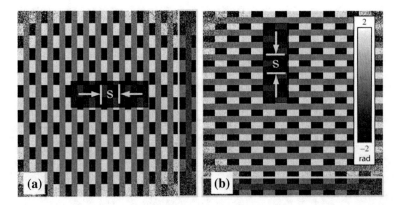

Fig. 7.17 Numerical simulation: phase difference based on 4 phase shifted shear interferograms with the shear set to **a** s_3 and **b** s_4. The shear is illustrated by the *arrow marks*. The *shaded region* bounded by the *box* indicates the area in which the values have been calculated using the natural extension Eq. (7.66)

bright fields. All distributions have a size of 510 by 510 sampling points. We simulated four phase shifted shear experiments with the shears set to $s_1 = (5,0)$, $s_2 = (0,5)$, $s_3 = (51,0)$ and $s_4 = (0,51)$ sampling points. For the phase shifting we used a four frame algorithm with 90° phase shift between the frames. The camera simulation comprised Poisson noise assuming a full well capacity of 500 electrons and an average dark current of 4 electrons.

In Fig. 7.17a, b we see the resulting noisy phase differences $m_{p,3}(r)$ and $m_{p,4}(r)$ of the simulated shear process for s_3 and s_4. The size and the orientation of the shears can be seen from the arrow marks. The values marked by the box and the shaded area have been calculated using the natural extension Eq. (7.66). Towards the edges we see increasing noise which arises from the reduced signal-to-noise ratio caused by the dominant Poisson noise in regions of low intensity on the phase shifted interference patterns.

The result of the inversion process using the Fourier approach based on Eq. (7.64) is shown in Fig. 7.18a. The checkerboard structure of the wavefront is clearly seen from the central region of the reconstruction. Towards the edges the noise level increases. This is a consequence of the low signal-to-noise ratio owing to the comparably low amplitude of the wavefield in these regions. In Fig. 7.18b the residual after subtraction of the known checkerboard structure is shown. To highlight the structure of the noise, the range of values is reduced by a factor of 5. Again we see the noise level increasing towards the corners with low SNR. Additionally, we can also see periodic distortions. This is typical for approaches operating in Fourier space. Since the inversion process is spatially independent, frequency dependent noise will be distributed across the entire spatial domain. As a consequence, we see shear-periodic artefacts in the reconstructed wavefront, because the signal-to-noise ratio of the shear process is frequency dependent. However, the standard deviation of the differences within the marked area is $\sigma = 0.035$ rad and,

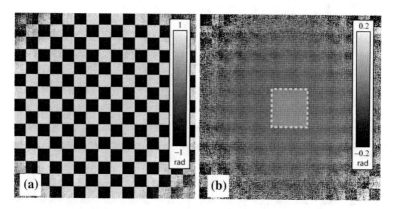

Fig. 7.18 Numerical simulation: **a** result obtained from the numerical inversion based on the Fourier approach Eq. (7.64) and **b** residual after subtraction of the known checkerboard structure of the phase. The standard deviation within the *marked area* is $\sigma = 0.035$ rad

therefore, still shows very good agreement between the initial and the reconstructed wavefront.

To avoid the periodic artefacts seen from Fig. 7.18b, we can introduce a noise dependent weighting through the weighting functions $w_n(r)$. As seen from Eq. (7.52), the phase shifting process provides us with the cross amplitudes $a(r)$ a $(r + s)$, which correspond to the modulation depth of the interference pattern. Hence, they can be regarded as a measure for the signal to-noise-ratio and we may set $w_n(r) = a(r) a(r + s_n)$. However, we cannot minimize Eq. (7.54) in the Fourier domain anymore, because of the non-linear characteristics of the weighting. Instead we will opt for an iterative non-linear optimization approach based on the steepest descent gradient method.

As mentioned before in Sect. 7.2.2, the basic idea of gradient search methods is to start with an initial guess, which is then iteratively improved by following the opposite direction of the gradient of the current estimate with respect to the parameters of interest. In our case we find for the successor

$$f^{(k+1)}(r) = f^{(k)}(r) - \alpha^{(k)} \cdot \nabla L^{(k)}(r), \qquad (7.67)$$

where the index k denotes the iteration and the scalar $\alpha^{(k)}$ is the step length of the kth iteration. By using Eq. (7.54) and employing the observations incorporating the noise terms $m_{n,\varepsilon}(r) = \Delta\varphi_n(r) + \varepsilon_n(r)$ rather than merely the phase differences $\Delta\varphi_n(r)$ we obtain the gradient

$$\nabla L^{(k)}(r) = \frac{\partial L}{\partial f^{(k)}(r)} = -2 \sum_n w_n(r) \Big[\Delta_n f^{(k)}(r) - m_{n,\varepsilon}(r) \\ + m_{n,\varepsilon}(r - s_n) - \Delta_n f^{(k)}(r - s_n) \Big], \qquad (7.68)$$

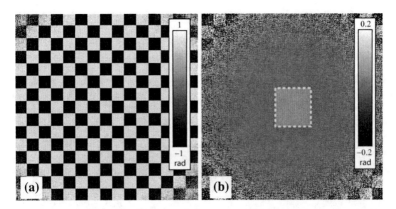

Fig. 7.19 Numerical simulation: **a** result obtained from the numerical inversion based on 200 iterations of the steepest descent gradient method Eq. (7.67) and **b** residual after subtraction of the known checkerboard structure of the phase. The standard deviation within the *marked area* is $\sigma = 0.017$ rad

where we have introduced $\Delta_n f(r) = f(r + s_n) - f(r)$ for brevity. The great benefit of this approach is that the gradient incorporates spatially resolved information about the measurement uncertainty which we have at hand anyway. We can demonstrate this advantage by looking at the reconstructed wavefront in Fig. 7.19a which was obtained after 200 iterations of Eq. (7.67). Visually it looks very similar to the one shown in Fig. 7.18a. However, if we look at the residual after subtraction of the known checkerboard structure in Fig. 7.19b we see that the periodic artefacts disappeared in those regions in which the wavefield has high intensities and the interference patterns consequently provide good signal-to-noise ratio. The standard deviation of the residual is now $\sigma = 0.017$ rad which is more than 3 dB better than the reconstruction based on the Fourier method. Additionally, pre-processing of the measured data by the natural extension is not necessary. The drawback of the method is the computational effort. It took a standard quad core CPU running at 2.53 GHz and equipped with 4 GB RAM approximately 1 min to perform the 200 iterations under MatLab.

Furthermore it is also possible to reconstruct the wavefront only across spatially limited pupil functions. This can be achieved by simply setting the weighting to zero outside the region of interest. An example is shown by Fig. 7.20a where the wavefront is reconstructed across two separated ring shaped pupils. Note though, that separation of individual regions is not a problem as long as neighboring regions overlap during at least one of the shear measurements. In Fig. 7.20b we see again the residual after subtraction of the known wavefront, showing no periodic distortions and no ringing at the edges.

Even if the objective function Eq. (7.54) is non-linear in the weighting, it is still possible to prove that any gradient based approach like the one given by Eq. (7.67) will eventually converge to a unique optimum solution in the least-squares sense. This follows from the fact that $L(f)$ is a convex function, which means that it only

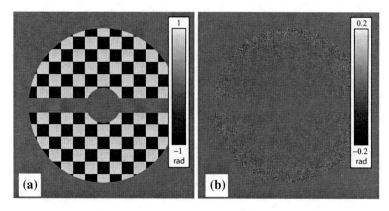

Fig. 7.20 Numerical simulation: **a** reconstruction of the wavefront across a pupil function based on 200 iterations of the steepest descent gradient method Eq. (7.67) and **b** residual after subtraction of the known checkerboard structure of the phase

exhibits a single global minimum. For a function to be convex it has to be ensured that the line segment connecting any two points of the function lies entirely above (or below) the function graph. This situation is depicted by Fig. 7.21. Formally, this requires for any of the potential solutions $f(\mathbf{r})$ and $g(\mathbf{r})$:

$$L(\alpha \cdot g + (1 - \alpha)f) \leq \alpha \cdot L(g) + (1 - \alpha)L(f), \qquad (7.69)$$

for any α between 0 and 1. It seems intuitively clear and straight forward to prove that a quadratic function $q(x;b) = (x - b)^2$ with constant shift b is convex and therefore satisfies

Fig. 7.21 Example of a convex function L: for any f, g and α, the *graph* of the function is below or equal to the line segment connecting $L(f)$ and $L(g)$

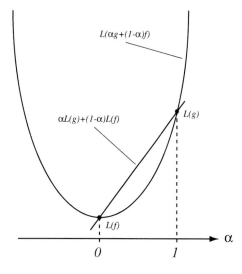

$$q(\alpha \cdot y + (1 - \alpha)x; b) \le \alpha \cdot q(y; b) + (1 - \alpha)q(x; b). \qquad (7.70)$$

To prove that Eq. (7.69) holds we will reduce it to the statement given by Eq. (7.70). Indeed we can reformulate the objective function Eq. (7.54) as a sum of functions, which are quadratic in $\Delta f(\mathbf{r})$:

$$L(f) = \sum_n \sum_{\mathbf{r}} q(\sqrt{w_n} \cdot \Delta_n f; b_n), \qquad (7.71)$$

where the constant is given by $b_n(\mathbf{r}) = w_n(\mathbf{r})^{0.5} m(\mathbf{r})$ and we have omitted the explicit dependence on $\mathbf{r}$ for brevity. It is also straight forward to show that

$$L(\alpha \cdot g + (1 - \alpha)f) = \sum_n \sum_{\mathbf{r}} q(\alpha\sqrt{w_n}\Delta_n g + (1 - \alpha)\sqrt{w_n}\Delta_n f; b_n). \qquad (7.72)$$

If we express $L(f)$ and similarly $L(g)$ by means of Eq. (7.71) and insert it together with Eq. (7.72) into Eq. (7.69), we can apply inequality Eq. (7.70) to any combination of n and $\mathbf{r}$ by substituting $x = w_n(\mathbf{r})^{0.5}\Delta_n f(\mathbf{r})$ and $y = w_n(\mathbf{r})^{0.5}\Delta_n g(\mathbf{r})$. Hence the inequality also holds for the entire sum, proving $L(f)$ to be convex.

Finally, we would like to present experimental results which demonstrate the remarkable low coherence requirements of shear interferometry. As seen from Eq. (7.53) it is sufficient that the mutual intensity $G(\mathbf{r}, \mathbf{r}')$ is significantly different from zero at any two points $\mathbf{r}$ and $\mathbf{r}' = \mathbf{r} + \mathbf{s}$ separated by the shear. This enables the application of a broader range of light sources, such as liquid crystal matrix displays (LCD) for example [56]. To verify this, we will first describe the intensity distribution $I_M(\mathbf{v})$ of light emitted by an LCD monitor as follows

$$I_M(\mathbf{v}) = \left[\mathrm{comb}\left(\frac{v_i}{\Delta p}, \frac{v_j}{\Delta p}\right) \otimes \mathrm{rect}\left(\frac{v_i}{p}, \frac{v_j}{p}\right) \right] \cdot \mathrm{rect}\left(\frac{v_i}{D}, \frac{v_j}{D}\right). \qquad (7.73)$$

Here, $\mathbf{v} = (v_i, v_j)^{\mathrm{T}}$ is a position vector, Δp is the pixel pitch, p is the size of the active area of a pixel, D is the size of the display, $\mathrm{comb}(\alpha, \beta) = \mathrm{comb}(\alpha)\mathrm{comb}(\beta)$ and $\mathrm{rect}(\alpha, \beta) = \mathrm{rect}(\alpha)\mathrm{rect}(\beta)$.

We can assume that the statistical properties of the emitted light are spatially stationary, i.e. do not vary in space. In this case $G(\mathbf{r}, \mathbf{r} + \mathbf{s}) = G(\mathbf{s})$ only depends on $\mathbf{s}$ and we can employ the *Van Cittert-Zernike theorem* to describe the relationship between the intensity distribution $I_M(\mathbf{v})$ and the mutual intensity $G(\mathbf{s})$ in any parallel plane at distance z, which has to be large compared to the extent of the light source

$$G(\mathbf{s}) = \iint I_M(\mathbf{v})\exp\left[i\frac{2\pi}{\lambda z}\mathbf{s} \cdot \mathbf{v} \right] dv_i dv_j. \qquad (7.74)$$

This is a scaled Fourier transform and we can insert Eq. (7.73) to obtain

$$G(s) = \left[\text{comb}\left(\frac{s_i \cdot \Delta p}{\lambda z}, \frac{s_j \cdot \Delta p}{\lambda z} \right) \text{sinc}\left(\frac{s_i \cdot p}{\lambda z}, \frac{s_j \cdot p}{\lambda z} \right) \right],$$
$$\otimes \text{sinc}\left(\frac{s_i \cdot D}{\lambda z}, \frac{s_j \cdot D}{\lambda z} \right)$$

$$\tag{7.75}$$

where $\text{sinc}(\alpha, \beta) = \text{sinc}(\alpha)\text{sinc}(\beta)$. By looking at the comb-function in Eq. (7.75) we see that $G(s)$ has a periodic structure and that it is indeed possible to perform shear interferometry if the shear s satisfies

$$s = \left(n\frac{\lambda z}{\Delta p}, m\frac{\lambda z}{\Delta p} \right)^T, \tag{7.76}$$

where n and m are integers. Having said this, we recognize that the first set of sinc-functions depending on p envelops the comb-function. Monitor pixels with small active areas are therefore preferable because the sinc-functions drop to zero for either $s_i = \lambda z/p$ or $s_j = \lambda z/p$. The worst case occurs when the pixel pitch equals the active area i.e. $\Delta p = p$, and the requirement Eq. (7.76) coincides with the zeroes of the sinc-functions. However, in this situation it is not pertinent to speak of a matrix display but rather of a spatially distributed light source of size D. The second set of sinc-functions broadens the individual spikes of the comb-function proportional to $\lambda z/D$, and therefore defines the degree of freedom within variation of the shear can be tolerated. Large variations, however, can only be realized with small light panels.

The shear-interferometer we used for the experiments is seen from Fig. 7.22. Similar to the setup for phase retrieval it is based on a 4f-configuration with a liquid crystal spatial light modulator (SLM) in the Fourier domain [55]. Here, we exploit

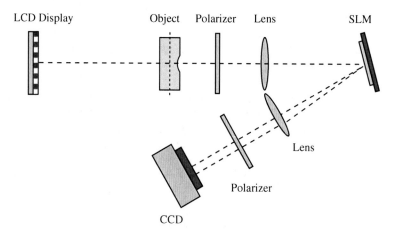

Fig. 7.22 An experimental setup for shear interferometry based on a spatial light modulator (*SLM*) in the spectral domain of a 4f-setup. For further details please refer to the text

the birefringent properties of an SLM, i.e. that light polarized along its slow axis will be diffracted while light polarized along the fast axis will be simply reflected from the back panel. We polarize the incident light exactly between the slow and the fast axis using the polarizer and let the SLM generate a blazed diffraction grating which exhibits a single diffraction order. Hence, two orthogonally polarized and mutually shifted images will appear in the sensor domain, where the shift can be electronically selected by the orientation and the period of the blazed grating generated by the SLM. The polarizing analyser in front of the camera is required to let the two images interfere. The main advantage of this configuration is its considerable tolerance against environmental disturbances and a fast, precise and highly reproducible adjustment of the shear. Additionally, the SLM can be used for temporal phase shifting.

The object under investigation was a 10 mm thick glass plate with refractive index of $n_{612} = 1.516$ and a parabolic indentation in its centre. The indentation has a diameter of 5 mm and a height of approximately 4 µm. The aim of the experiment is to determine the exact height of the deformation by measuring the optical path of light traveling through the object. The LCD matrix display of an Apple iPhone 4S is used as a light source at a distance of 68 mm from the object. The pixel pitch of the LCD is approximately 78 µm. However, for the sake of light efficiency we combined 2 by 2 LCD pixels creating a periodic illumination with a respective active area of size $p = 156$ µm and a pixel pitch of $\Delta p = 234$ µm. We only used red pixels which emit light at a central wavelength of $\lambda = 612$ nm. The size of the periodic illumination was $D = 15$ mm in any direction giving us only a small degree of freedom to vary the shear. We performed 4 phase shifting shear experiments $M_1(r) - M_4(r)$ with the shears set to 50 and 51 camera pixels in horizontal and vertical direction respectively, where the camera has a pixel pitch of 3.45 µm.

Figure 7.23a, b show by example the measured amplitude and phase difference for the shear set to 50 sensor pixels in horizontal direction. In regions were the objects surface exhibits steep gradients we see that the amplitude is comparably

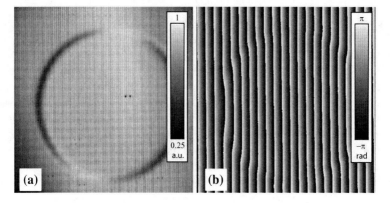

Fig. 7.23 Example of a phase shifted measurement with the shear set to 50 sensor pixels in *horizontal direction*: **a** amplitude and **b** phase of the cross term $M_1(r)$ [see Eq. (7.52)]

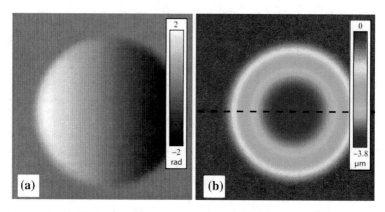

Fig. 7.24 Experimental results: **a** the phase of $M_1(r)$ after subtraction of the linear fringes caused by the spherical illumination *waves* and **b** the shape of the indention calculated from the reconstructed the wavefront by means of Eq. (7.77)

low, which demonstrates the influence of the objects shape on the mutual intensity in the sensor domain. Furthermore, we see periodic distortions which indicate a residual spatial dependence of the mutual intensity owing to the fact that the light source is relatively close to the object plane. The measured phase differences in Fig. 7.23b are dominated by straight fringes which arise from the spherical waves emerging from each of the LCD pixels. Since this is an artefact of the illumination we performed a linear fit with coefficients taken from regions assumed to be flat in vicinity to the indention. The result is seen from Fig. 7.24a. The distributions corresponding to all four shears have been subjected to the iterative reconstruction scheme constituted by Eq. (7.67). From the resulting estimate $f^{(k)}$ of the phase it is possible to calculate the height of the object by

$$h(r) = \frac{\lambda}{2\pi(n_{612} - n_0)} f^{(k)}(r), \qquad (7.77)$$

where the refractive index of air can be assumed $n_0 = 1$.

The result for $k = 2{,}000$ iterations can be seen from Fig. 7.24b. The profile indicates a depth of the indention of $d_S = 3.72$ μm (peak-to-valley) along the dashed line. This shows very good agreement with the result of a comparison measurement at the *Physikalisch Technische Bundesanstalt* (PTB) using a calibrated Fizeau interferometer, which yielded $d_F = 3.67$ μm. The standard deviation of the wavefront estimation is $\sigma = 0.008\lambda$, where the largest deviations are caused by a periodic pattern arising from the residual spatial dependence of the complex mutual intensity seen from Fig. 7.23a. This example shows the remarkably low demands of shear interferometry with regard to coherence requirements. It is possible to achieve interferometric accuracy even when light of an LCD monitor is used.

7.3.2 *Computational Shear Interferometry*

In the preceding section we have presented methods to recover a smooth wavefront from measurements based on a shear interferometer. However, in many applications it is favorable to determine the entire complex amplitude of a monochromatic wavefield. For example, in all cases in which diffracted light or speckle fields are investigated, it cannot be assumed that the corresponding wavefront is smooth. Indeed there are only specific situations in which the smoothness assumption is valid.

In the following we will refer to methods which aim at recovering the full complex amplitude of monochromatic wavefields by means of a shear interferometer as computational shear interferometry (CoSI). In adaptive optics, Fried has reported on a heuristic method to recover the complex amplitude of a wavefield [66]. It works on square pupil functions with a size of $2^N + 1$ times $2^N + 1$ samples and requires two orthogonal shears with magnitude of a single pixel oriented along the main axes of the sampling device. For standard interferometric applications, such as speckle interferometry or quantitative phase imaging it is probably more convenient to directly minimize Eq. (7.55), for example by means of gradient based techniques [57]. This has the advantage that the orientation and the magnitude of the shears can be freely chosen as long as the uniqueness of the solution is ensured, that the inversion can be performed on arbitrarily shaped pupil functions and that additional pre-knowledge can be easily added by means of regularization. In this section, we will therefore discuss an approach based on the steepest descent gradient method and give some experimental examples demonstrating the capabilities of CoSI.

We have used gradient based approaches already in Sects. 7.2.2 and 7.3.1. The basic iterative scheme is well represented by Eq. (7.67) with the difference that the estimates $f^{(k)}(r)$ are complex valued amplitudes rather than scalar phase values and that the gradient has to be derived from Eq. (7.55). The latter is not a straight forward task, because $L(f)$ is not an analytic complex function, i.e. it is not holomorphic, but rather a real valued function which depends on the complex variable f $(r) = f_R(r) + if_I(r)$, exhibiting a real part $f_R(r)$ and an imaginary part $f_I(r)$. As a consequence, the derivative

$$\frac{dL}{df} = \lim_{\Delta f \to 0} \frac{L(f + \Delta f) - L(f)}{\Delta f}, \tag{7.78}$$

does not exist, because it depends on the direction from which Δf approaches zero. In this unfortunate situation we may remember that the intention behind calculation of the gradient was to determine the variation of the value of L with respect to variation of the parameter f. Having this in mind, we may define

$$\nabla L = \frac{dL}{df} = \frac{\partial L}{\partial f_R} + i \frac{\partial L}{\partial f_I}, \qquad (7.79)$$

as a helpful tool to substitute for the non-existent gradient. Please note that according to Eq. (7.79) for $g(z) = z$ we find $dg/dz = 0$. It is therefore not an admissible definition of a complex derivative but rather provides a mapping of the complex plane onto two dimensional Cartesian coordinates. However, the partial derivatives make it feasible to define a direction in the complex plane along which the value of L increments most and we may use it in Eq. (7.67). Inserting Eq. (7.55) into Eq. (7.79) yields

$$\nabla L^{(k)}(\mathbf{r}) = -2 \sum_n f^{(k)}(\mathbf{r} + \mathbf{s}_n) \psi_n^{(k)*}(\mathbf{r}) + f^{(k)}(\mathbf{r} - \mathbf{s}_n) \psi_n^{(k)}(\mathbf{r} - \mathbf{s}_n), \qquad (7.80)$$

where

$$\psi_n^{(k)}(\mathbf{r}) = M_n(\mathbf{r}) - f^{(k)*}(\mathbf{r}) f^{(k)}(\mathbf{r} + \mathbf{s}_n), \qquad (7.81)$$

and $M_n(\mathbf{r})$ according to Eq. (7.52). To discuss the uniqueness of solutions to Eq. (7.55) we will set the gradient to zero and arrive at an implicit definition of the solution

$$f(\mathbf{r}) = \frac{\sum_n f(\mathbf{r} + \mathbf{s}_n) \cdot M_n^*(\mathbf{r}) + f(\mathbf{r} - \mathbf{s}_n) \cdot M_n(\mathbf{r} - \mathbf{s}_n)}{\sum_n |f(\mathbf{r} - \mathbf{s}_n)|^2 + |f(\mathbf{r} + \mathbf{s}_n)|^2}, \qquad (7.82)$$

In case of a single measurement $n = 1$ we see that if $f(\mathbf{r})$ is a solution, then any $g(\mathbf{r}) = f(\mathbf{r}) \exp[i\phi_p(\mathbf{r})]$ is a solution as well, where $\phi_p(\mathbf{r})$ is an arbitrary shear periodic function $\phi_p(\mathbf{r} + \mathbf{s}) = \phi_p(\mathbf{r})$. Furthermore if $f(\mathbf{r})$ is a solution, then any $g(\mathbf{r}) = a_p(\mathbf{r}) f(\mathbf{r})$ is a solution as well, where $a_p(\mathbf{r})$ is a periodic real valued function for which $a_p(\mathbf{r} + \mathbf{s}) = 1/a_p(\mathbf{r})$ holds. We see that, similar to that in wavefront reconstruction (Sect. 7.3.1), the inversion is not unique when only a single shear is considered. Again, the unambiguity is related to the period of the shear and we may ensure the uniqueness of the solution by evaluating the results of several measurements with varying orientations and magnitudes of the shear in combination. Also here, a good choice is to select orthogonal shears of which the magnitudes are relatively prime.

A major difference between wavefront reconstruction and wavefield reconstruction in computational shear interferometry is that the objective function Eq. (7.55) is not convex. Hence it has multiple local minima in which any gradient based approach can become trapped. At which of the stationary points the algorithm eventually arrives depends on the initial guess $f^{(0)}(\mathbf{r})$. A Kronecker delta distribution has been shown to be a good initial guess

$$f^{(0)}(r) = \begin{cases} 1, & r = r_0 \\ 0, & \text{elsewhere} \end{cases}. \tag{7.83}$$

The basic functionality can be best understood from looking at the form of the gradient in Eq. (7.80). It can be seen that only those parts of the current estimate $f^{(k)}(r)$ will be changed during the iteration which have at least one nonzero value in their neighbourhood, where the neighbourhood is defined by the shear. This shows that the choice of a Kronecker delta as an initial guess starts the reconstruction process at point r_0 and progresses in outward direction for consecutive iterations. The procedure behaves as a path-dependent approach, which walks all available paths at the same time and additionally balances accumulated errors between them. Even though we cannot prove it formally, we conjecture in many cases from simulations and comparison measurements with standard interferometry that the approach finds the global minimum, or at least a solution very close to it.

In addition, we can add further pre-knowledge to the optimization process. A simple example is to constraint the solution to a smooth function by means of regularization. Please note that the complex amplitude is a solution to the Helmholtz-Equation and therefore is always smooth even if the corresponding wavefront is not. Smoothness can be constraint by adding a minimum curvature term to the objective function

$$L(f) = \sum_n \|M_n(r) - f^*(r)f(r + s_n)\|^2 + \gamma\|\Delta_D f(r)\|^2. \tag{7.84}$$

Here, γ is the regularization parameter. Large values of γ will yield very smooth estimates of the wavefield. The discrete Laplacian Δ_D for a two dimensional discrete function $g(x,y)$ is given by

$$\begin{aligned} \Delta_D g(x,y) = & \, g(x+1,y) + g(x-1,y) \\ & + g(x,y+1) + g(x,y-1) - 4g(x,y). \end{aligned} \tag{7.85}$$

The corresponding gradient is calculated by inserting Eq. (7.84) into Eq. (7.89) which yields

$$\nabla L_S^{(k)}(r) = \nabla L^{(k)}(r) + 2\gamma\Delta_D\left\{\Delta_D f^{(k)}(r)\right\}. \tag{7.86}$$

The setup shown in Fig. 7.25 was used to demonstrate the potential of computational shear interferometry for wavefield sensing. On the right hand side we see an SLM-based shear interferometer, similar to the one introduced in Fig. 7.22. The object is a dice with an edge length of 8 mm as seen from the detail. It is positioned in a distance of $d = 130$ mm from the input plane of the shear interferometer. The aim of the experiment is to measure the complex amplitude of the wavefield scattered by the object. Shear interferograms were recorded with the shears set to 17 and 20 camera pixels in the horizontal and vertical directions, respectively.

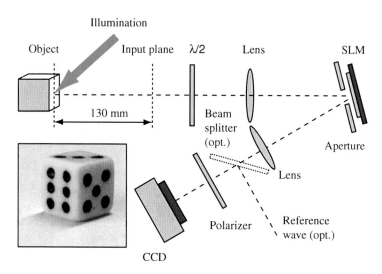

Fig. 7.25 An experimental setup to demonstrate computational shear interferometry: the aim of the experiment is to determine the complex amplitude of light scattered by the dice object using shear interferometry. The beam splitter can be used to optionally superpose the wavefield in the sensor domain by a plane reference wave in order to compare the results of CoSI with those of standard phase shifting interferometry. For further details on the setup please refer to the text

A four-frame 90° phase-shifting algorithm was used to determine the observations $M_1(r) - M_4(r)$ from a total of 16 recorded interferograms.

In order to compare the results with those obtained from standard interferometry, a beam splitter can be used to superpose the wavefield in the sensor domain with a plane reference wave. In this configuration, the polarization is selected towards the slow axis of the SLM using the half-wave plate, thus turning it into a phase shifting device. The light source is a laser emitting light at $\lambda = 532$ nm.

In Fig. 7.26a, b we see amplitude and phase, respectively, of the wavefield across the input plane of the shear interferometer as obtained after 150 iterations of Eq. (7.67) with the gradient Eq. (7.86), $\gamma = 2$ and a Kronecker delta as initial guess. As seen from the detail in Fig. 7.26b, the phase distribution does not constitute a smooth wavefront but rather exhibits a large number of phase singularities.

In Fig. 7.27a the difference between the phase distribution obtained from computational shear interferometry and standard phase shifting interferometry is shown. The standard deviation is $\sigma = 0.68$ rad, indicating very good agreement. Similar to what is done in phase shifting digital holography we can use the wavefield so found to reconstruct the object by numerical propagation into the object plane. The result is depicted by Fig. 7.27b where the object is clearly seen to be in focus.

The low demands on coherence make it feasible to record digital phase shifted holograms of rough objects using LED illumination. This can be seen from Fig. 7.28a, b, where the amplitude and the phase, respectively, of light scattered by

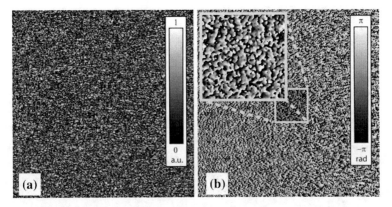

Fig. 7.26 Reconstructed complex amplitude after 150 iterations: **a** normalized amplitude and **b** phase of the light scattered by the *dice*. The size of the distributions is 680 × 680 sensor pixels, with a pixel size of $\Delta p = 3.45$ μm. This result is comparable to a phase shifted digital hologram

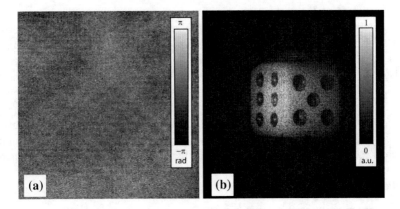

Fig. 7.27 **a** Difference between the phase distributions obtained from CoSI and standard phase shifting interferometry. The standard deviation is $\sigma = 0.68$ rad which shows very good agreement. None of the singularities has been wrongly detected by the iterative algorithm. **b** Resulting intensity after numerical propagation of the wavefield by $d = 130$ mm in the object plane. The object is clearly seen to be in focus, similar to what is expected from the reconstruction of a digital hologram

a 1 cent Euro coin is shown. The light source was a fibre coupled LED with a coherence length of $l_c = 10$ μm and a fibre diameter of $d_f = 200$ μm. Again, four phase shifted measurements with the shears selected to 3 and 5 pixels in both, horizontal and vertical direction were evaluated in combination. To arrive at the shown complex amplitude $k = 950$ iterations of Eq. (7.67) with gradient Eq. (7.86) and $\gamma = 2$ were required. The object was positioned approximately 5 mm away from the front focal plane of the shear interferometer. Hence the reconstructed amplitude in Fig. 7.28a appears to be a blurred representation of the coins surface. To verify

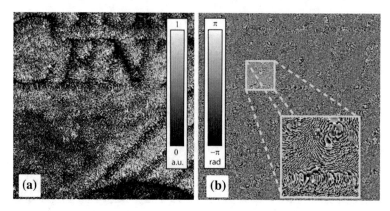

Fig. 7.28 The complex amplitude of light scattered by a coin under LED illumination measured by means of computational shear interferometry: **a** normalized amplitude and **b** phase of the light. The size of the distributions is 700 × 700 sensor pixels, with a pixel size of $\Delta p = 6.9$ μm

the correctness of the result, the intensity after propagation of the wavefield by 5.3 mm towards the object plane is seen in Fig. 7.29a, showing the object clearly in focus. For comparison, Fig. 7.29b depicts a photo of the object with the investigated region of the surface marked by the box.

The above results indicate that CoSI can be used in a large number of situations in which phase shifting interferometry or phase-shifting digital holography are applied. Yet no reference wave is required which offers all the benefits obtained from a common path approach, such as robustness against environmental disturbances and comparably low demands regarding the spatial and the temporal coherence of the investigated wavefield. In contrast to phase retrieval no inherent diversity is required so that smooth wavefronts and speckle fields can be measured by the same method.

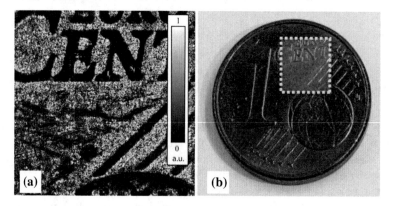

Fig. 7.29 Object in focus: **a** intensity distribution across the object plane as obtained after numerical propagation of the complex amplitude in Fig. 7.28 by 5.3 mm and **b** the object with the investigated region *marked by the box*

7.4 Shack-Hartmann Wavefront Sensing

Shack-Hartmann [80, 212] sensing is mainly used to determine smooth wavefronts. This is done by means of a lens array in front of a camera sensor, where the distance between the sensor and the array is given by the focal length f of the lenses. This configuration can be seen from Fig. 7.30. The angles $\alpha(r)$ and $\beta(r)$ that the wavefront includes with the axes of the sensing device depend on its local slope. This causes a shift d of the respective focal spot according to

$$\mathbf{d}(r) = f \begin{pmatrix} \cot \alpha(r) \\ \cot \beta(r) \end{pmatrix}. \tag{7.87}$$

The shifts are proportional to the local phase gradient

$$\nabla \phi(r) = \frac{2\pi}{\lambda} \begin{pmatrix} \cos \alpha(r) \\ \cos \beta(r) \end{pmatrix} \approx \frac{2\pi}{\lambda f} \mathbf{d}(r), \tag{7.88}$$

where we have used the approximation $\cos(\alpha) \approx \cot(\alpha)$ for $\alpha \approx 90°$, which means small angles of the incident light with respect to the optical axes of the lenses. Similar to what we have done in phase retrieval and in shear interferometry, we may recover the wavefront by numerical integration in the Fourier domain. The transfer function of differentiation can be deduced from Eqs. (7.40) and (7.41) and is given by $H_x(v,\xi) = i2\pi v$ and $H_y(v,\xi) = i2\pi\xi$. For integration, we can reformulate the minimum norm least square approach constituted by Eq. (7.64)

$$F(v,\xi) = \frac{M_{x,\varepsilon}(v,\xi) \cdot H_x^*(v,\xi) + M_{y,\varepsilon}(v,\xi) \cdot H_y^*(v,\xi)}{|H_x(v,\xi)|^2 + |H_y(v,\xi)|^2 + \alpha} \tag{7.89}$$

Fig. 7.30 Principle of Shack-Hartmann wavefront sensing: the slope of the wavefront shifts the focal spots of the lens array. The shift is detected by a camera sensor and is proportional to the phase gradient

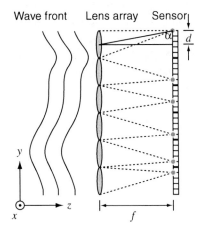

where $M_{x,\varepsilon}(v,\xi) = F\{\nabla\phi_x(r) + \varepsilon_x(r)\}$ and $M_{y,\varepsilon}(v,\xi) = F\{\nabla\phi_y(r) + \varepsilon_y(r)\}$ are the Fourier transforms of the observations of the gradient in x and y direction respectively, and $\varepsilon_x(r)$ and $\varepsilon_y(r)$ describe additive stationary noise. From $F(v,\xi)$ the wavefront $f(r) = F^{-1}\{F(v,\xi)\}$ can be obtained by an inverse Fourier transform.

When it can be assumed that the wavefront under investigation is rotationally symmetric or merely shows comparably small variations, it is common practice to derive an optimum set of Zernike coefficients from the measured slopes [188]. If phase singularities are present, the phase distribution of the wavefield can no longer be described by a smooth function. In this case, similar to what we have discussed in computational shear interferometry, more sophisticated iterative techniques have to be applied [14]. Finally we would like to mention that evaluation of the intensity across the focal spots also allows estimating the corresponding amplitude and therefore facilitates the determination of the full complex amplitude of the underlying wavefield [160].

The major advantage of Shack-Hartmann sensors is that they are very fast. They only require a single camera frame to estimate the wavefront and are widely used to determine the wavefront slopes of light travelling through the atmosphere. The measured data is used as a control in adaptive optics which compensate for distortions caused by turbulence thereby enhancing the seeing of earth bound telescopes. Having said this, a major disadvantage is the low space-bandwidth-product due to the fact that a large number of camera pixels (typically 10 by 10) have to be sacrificed for one sampling point of the Shack-Hartmann sensor. As a consequence, common devices only provide 100 by 100 sampling points. Furthermore, the maximum angle is limited because of the imaging properties of the lenses and in order to avoid cross-talk between neighbouring sampling points. Over the past few years, strategies have been developed to compensate for these drawbacks. Seifert et al. suggest creating an array of virtual lenses by means of a liquid crystal spatial light modulator (SLM) instead of a fixed lens array [209]. The focal length of the virtual lens array can be actively controlled thereby enhancing the dynamic range of the sensor with respect to the maximum angle. A similar approach is reported by Hongbin et al. [87] using an array of fluidic micro lenses instead of an SLM. To compensate for lens aberrations under large incidence angles Grunwald et al. [76] suggest using an array of micro-axicons instead of micro lenses. The corresponding sensor enables detection of wavefront slopes within a range $\pm30°$ which is currently the technical limit.

Chapter 8
Speckle Metrology

8.1 Electronic Speckle Pattern Interferometry (ESPI)

Electronic Speckle Pattern Interferometry (ESPI) is a method, similar HI, to measure optical path changes caused by deformation of opaque bodies or refractive index variations within transparent media [82, 147]. In ESPI electronic devices (CCD or CMOS) are used to record the information. The speckle patterns which are recorded by an ESPI system can be considered as holograms of focused images and are formed in the image plane. Due to the digital recording and processing, ESPI is also known as *Digital Speckle Pattern Interferometry* (DSPI) or *TV-holography*. However, instead of hologram reconstruction the speckle patterns are correlated.

The principal set-up of an Electronic Speckle Pattern Interferometer is shown in Fig. 8.1. The object is imaged onto a camera by a lens system. Due to the coherent illumination the image formed is a speckle pattern. According to Eq. (2.57) the speckle size depends on the wavelength, the image distance and the aperture diameter. For good imaging, the speckle size should be of the same order as the pixel size (resolution) of the electronic sensor. This can be achieved by closing the aperture of the imaging system.

The speckle pattern of the object surface is superimposed on the target with a spherical reference wave. The source point of the reference wave should be located in the centre of the imaging lens. Due to this in-line configuration the spatial frequencies are resolvable by the sensor. In practice the reference wave is coupled into the set-up by a beam splitter (as shown in Fig. 8.1) or guided via an optical fibre, which is mounted directly in the aperture of the lens system.

The intensity at the target is:

$$I_A(x, y) = |a_R(x, y)\exp(i\varphi_R) + a_O(x, y)\exp(i\varphi_O)|^2$$
$$= a_R^2 + a_O^2 + 2a_R a_O \cos(\varphi_O - \varphi_R) \tag{8.1}$$

© Springer-Verlag Berlin Heidelberg 2015
U. Schnars et al., *Digital Holography and Wavefront Sensing*,
DOI 10.1007/978-3-662-44693-5_8

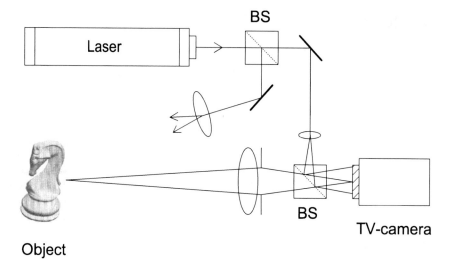

Fig. 8.1 Electronic speckle pattern interferometer

where $a_R \exp(i\varphi_R)$ is the complex amplitude of the reference wave and $a_O \exp(i\varphi_O)$ is the complex amplitude of the object wave in the image plane. The term $(\varphi_O - \varphi_R)$ is the phase difference between reference and object wave, which varies randomly from point to point. This speckle interferogram is recorded and electronically stored.

The set-up in Fig. 8.1 is sensitive to out-of-plane deformations, i.e. deformations perpendicular to the object surface. A displacement of d_z corresponds to a phase shift of

$$\Delta\varphi = \frac{4\pi}{\lambda} d_z \qquad (8.2)$$

After deformation a second speckle pattern is recorded:

$$\begin{aligned}
I_B(x,y) &= |a_R(x,y)\exp(i\varphi_R) + a_O(x,y)\exp(i\varphi_O + \Delta\varphi)|^2 \\
&= a_R^2 + a_O^2 + 2a_R a_O \cos(\varphi_O - \varphi_R + \Delta\varphi)
\end{aligned} \qquad (8.3)$$

These two speckle pattern are now subtracted:

$$\begin{aligned}
\Delta I &= |I_A - I_B| = |2a_R a_O (\cos(\varphi_O - \varphi_R) - \cos(\varphi_O - \varphi_R + \Delta\varphi))| \\
&= 2a_R a_O \left| \sin\left(\varphi_O - \varphi_R + \frac{\Delta\varphi}{2}\right) \sin\frac{\Delta\varphi}{2} \right|
\end{aligned} \qquad (8.4)$$

The intensity of this difference image is minimal at those positions, where $\Delta\varphi = 0, 2\pi, \ldots$. The intensity reaches its maximum at those positions, where

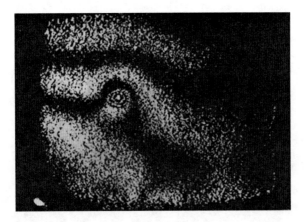

Fig. 8.2 ESPI image

$\Delta\varphi = \pi, 3\pi, \ldots$. The result is a pattern of dark and bright fringes, similar to a holographic interferogram. However, when compared to HI three-dimensional information in the correlation process is lost. Also the presence of speckle gives the fringes a more granular appearance A typical ESPI subtraction pattern is shown in Fig. 8.2.

As already mentioned, the set-up of Fig. 8.1 is only sensitive to out-of plane motion. In-plane displacements can be measured using the arrangement of Fig. 8.3. Two plane waves illuminate the object symmetrically at the angles $\pm\theta$ to the z-axis. The object is imaged by a camera. Again the speckle size is adapted to the target resolution by the aperture of the imaging system. The phase change due to an in-plane displacement can be derived by geometrical considerations, similar to the HI displacement calculations. The phase change of the upper beam is

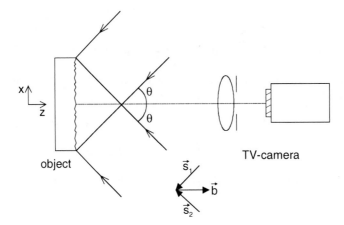

Fig. 8.3 In-plane sensitive speckle interferometer

$$\Delta \varphi_1 = \frac{2\pi}{\lambda} \vec{d} \left(\vec{b} - \vec{s}_1 \right) \qquad (8.5)$$

with displacement vector $\vec{d}$. The unit vectors $\vec{b}$, $\vec{s}_1$ and $\vec{s}_2$ are defined in Fig. 8.3. The corresponding phase shift of the lower beam is

$$\Delta \varphi_2 = \frac{2\pi}{\lambda} \vec{d} \left(\vec{b} - \vec{s}_2 \right) \qquad (8.6)$$

The total phase shift is

$$\Delta \varphi = \Delta \varphi_1 - \Delta \varphi_1 = \frac{2\pi}{\lambda} \vec{d} (\vec{s}_2 - \vec{s}_1) \qquad (8.7)$$

The vector $(\vec{s}_2 - \vec{s}_1)$ is parallel to the x-axis, its length is $2 \sin \theta$. The result for the total phase shift as measured by the camera is therefore:

$$\Delta \varphi = \frac{4\pi}{\lambda} d_x \sin \theta \qquad (8.8)$$

By using non-symmetrical illumination, the method also becomes sensitive to out-of-plane displacements.

As for HI, the phase cannot be determined from a single speckle pattern. The interference phase has to be recovered by, phase shifting methods [35, 223, 224]. Phase shifting ESPI requires recording at least three speckle interferograms with mutual phase shifts *in each state*. Any of the various phase shifting methods can be applied. Here an algorithm with 4 recordings and an unknown, but constant phase shift angle α, is used. The equations representing the initial state are:

$$\begin{aligned}
I_{A,1} &= a_R^2 + a_O^2 + 2a_R a_O \cos(\varphi_0 - \varphi_R) \\
I_{A,2} &= a_R^2 + a_O^2 + 2a_R a_O \cos(\varphi_0 - \varphi_R + \alpha) \\
I_{A,3} &= a_R^2 + a_O^2 + 2a_R a_O \cos(\varphi_0 - \varphi_R + 2\alpha) \\
I_{A,4} &= a_R^2 + a_O^2 + 2a_R a_O \cos(\varphi_0 - \varphi_R + 3\alpha)
\end{aligned} \qquad (8.9)$$

The dependence of the intensities and amplitudes from the spatial coordinates (x, y) has been omitted. This equation system has following solution:

$$\varphi_0 - \varphi_R = \arctan \frac{\sqrt{I_{A,1} + I_{A,2} - I_{A,3} - I_{A,4}} \cdot \sqrt{3I_{A,2} - 3I_{A,3} - I_{A,1} + I_{A,4}}}{I_{A,2} + I_{A,3} - I_{A,1} - I_{A,4}} \qquad (8.10)$$

For the second state 4 phase shifted interferograms are also recorded:

$$I_{B,1} = a_R^2 + a_O^2 + 2a_R a_O \cos(\varphi_0 - \varphi_R + \Delta\varphi)$$
$$I_{B,2} = a_R^2 + a_O^2 + 2a_R a_O \cos(\varphi_0 - \varphi_R + \Delta\varphi + \alpha)$$
$$I_{B,3} = a_R^2 + a_O^2 + 2a_R a_O \cos(\varphi_0 - \varphi_R + \Delta\varphi + 2\alpha)$$
$$I_{B,4} = a_R^2 + a_O^2 + 2a_R a_O \cos(\varphi_0 - \varphi_R + \Delta\varphi + 3\alpha)$$

$$(8.11)$$

The solution is:

$$\varphi_0 - \varphi_R + \Delta\varphi = \arctan \frac{\sqrt{I_{B,1} + I_{B,2} - I_{B,3} - I_{B,4}} \cdot \sqrt{3I_{B,2} - 3I_{B,3} - I_{B,1} + I_{B,4}}}{I_{B,2} + I_{B,3} - I_{B,1} - I_{b,4}}$$

$$(8.12)$$

The interference phase $\Delta\varphi$ is now calculated from Eqs. (8.10) and (8.12) by subtraction.

Phase shifting speckle interferometry is sometimes also called *Electro-Optic Holography* (EOH).

8.2 Digital Shearography

ESPI as well as conventional and Digital HI are highly sensitive to optical path changes. Displacement measurements up to a resolution of $\lambda/100$ are possible. On the other hand this high sensitivity is also a drawback for applications in field environments, where vibration isolation may not be available. Unwanted optical path length variations due to vibrations disturb the recording process.

As we have already seen in Chap. 7, *Shearography* [15, 89, 90, 170] is an interferometric method, which brings the rays scattered from one point of the object $P(x, y)$ into interference with those from a neighbouring point $P(x + \Delta x, y)$. The distance between both points is Δx. The shearing can be realized by mounting a glass wedge, in one half of the imaging system, Fig. 8.4. The object is imaged via both halves of the aperture (with and without wedge). Therefore two laterally sheared images overlap at the recording device, see Fig. 8.5.

Fig. 8.4 Speckle shearing interferometer

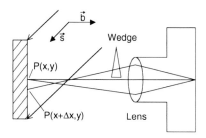

Fig. 8.5 Image from a
shearing camera

The intensity on the target is:

$$I_A(x, y) = |a_1(x, y) \exp(i\varphi(x, y)) + a_2(x, y) \exp(i\varphi(x + \Delta x, y))|^2$$
$$= a_1^2 + a_2^2 + 2a_1a_2 \cos(\varphi(x, y) - \varphi(x + \Delta x, y)) \tag{8.13}$$

where $a_1 \exp(i\varphi(x, y))$ and $a_2 \exp(i\varphi(x + \Delta x, y))$ are the complex amplitudes of the
interfering waves in the image plane. As in ESPI the phase difference
$(\varphi(x + \Delta x, y) - \varphi(x, y))$ varies randomly from point to point. This speckle interferogram is recorded and electronically stored. Another interferogram is recorded
for the second state B:

$$I_B(x, y) = \begin{vmatrix} a_1(x, y) \exp[i(\varphi(x, y) + \Delta\varphi(x, y))] \\ + a_2(x, y) \exp[i(\varphi(x + \Delta x, y) + \Delta\varphi(x + \Delta x, y))] \end{vmatrix}^2$$
$$= a_1^2 + a_2^2 + 2a_1a_2 \cos[\varphi(x, y) - \varphi(x + \Delta x, y) + \Delta\varphi(x, y) - \Delta\varphi(x + \Delta x, y)]$$
$$\tag{8.14}$$

Pointwise subtraction gives:

$$\Delta I = |I_A - I_B|$$
$$= \left| 2a_1a_2 \left\{ \cos(\varphi(x, y) - \varphi(x + \Delta x, y)) - \cos\begin{bmatrix} \varphi(x, y) - \varphi(x + \Delta x, y) \\ + \Delta\varphi(x, y) - \Delta\varphi(x + \Delta x, y) \end{bmatrix} \right\} \right|$$
$$= 2a_1a_2 \left| \sin\left\{ \varphi(x, y) - \varphi(x + \Delta x, y) + \frac{\Delta\varphi(x,y) - \Delta\varphi(x+\Delta x,y)}{2} \right\} \right.$$
$$\left. \times \sin\frac{\Delta\varphi(x,y) - \Delta\varphi(x+\Delta x,y)}{2} \right|$$
$$\tag{8.15}$$

This correlation pattern is known as a *shearogram*, see typical example in
Fig. 8.6.

Fig. 8.6 Shearogram

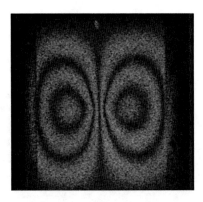

The phase shift due to deformation in the argument of Eq. (8.15) is calculated as follows (see also the definition of unit vectors $\vec{b}$ and $\vec{s}$ in Fig. 8.4):

$$\Delta\varphi(x,y) - \Delta\varphi(x + \Delta x, y) = \frac{2\pi}{\lambda}\left\{\vec{d}(x,y)\left(\vec{b} - \vec{s}\right) - \vec{d}(x + \Delta x, y)\left(\vec{b} - \vec{s}\right)\right\}$$

$$= \frac{2\pi}{\lambda}\left\{\frac{\vec{d}(x,y) - \vec{d}(x + \Delta x, y)}{\Delta x}\left(\vec{b} - \vec{s}\right)\right\}\Delta x$$

$$\approx \frac{2\pi}{\lambda}\frac{\partial\vec{d}(x,y)}{\partial x}\left(\vec{b} - \vec{s}\right)\Delta x$$

$$(8.16)$$

A shearing interferometer is therefore sensitive to the derivative of the displacement into the shear direction, in contrast to ESPI which is sensitive to the displacement. Shearography is relatively insensitive for rigid body motions, because $\partial\vec{d}(x,y)/\partial x$ vanishes if the object is moved as a whole [219, 220]. A second property which makes a shearing interferometer less sensitive to vibrations is the self-reference principle: Optical path changes due to vibrations influence both partial beams, which means they compensate each other to a certain degree. Shearography is therefore suited for rough environments with low vibration isolation.

The measurement sensitivity of a speckle shearing interferometer can be adjusted by varying the magnitude of the shear Δx. This parameter is determined by the wedge angle in the interferometer set-up of Fig. 8.4. Other shearing interferometer geometries are based on a Michelson interferometer, where the mirror tilt determines the shearing, Fig. 8.7.

Phase shifting techniques can also be applied in shearography. As with ESPI a set of phased shifted images is recorded in each state from which the phase according to Eq. (8.16) is calculated, see example in Fig. 8.8. An aircraft fuselage damaged in a hailstorm is investigated by shearography. The figure shows a filtered mod 2π phase image.

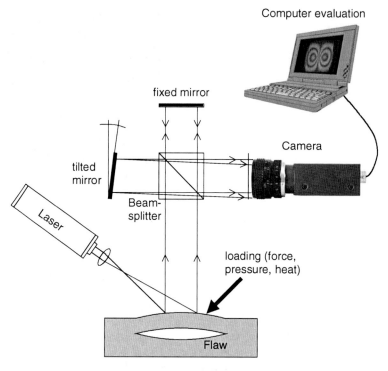

Fig. 8.7 Shearography set-up based on a Michelson interferometer

Fig. 8.8 Shearography phase
image. Courtesy of M. Kalms

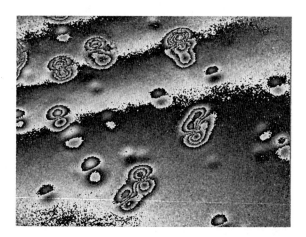

8.3 Digital Speckle Photography

Digital Speckle Photography (DSP) is the electronic version of Speckle Photography [29, 86, 134, 215, 230]. The method is used to measure in-plane displacement and strain. In classical Speckle Photography, two speckle patterns of the same surface are recorded on photographic film, e.g. with the set-up of Fig. 2.11. The object suffers an in-plane deformation between the exposures. This in-plane deformation is made visible as fringe pattern by pointwise illumination of the double exposed film with a collimated laser beam or alternatively using an optical filtering set-up. In DSP the speckle patterns are recorded by a high resolution electronic sensor camera, electronically stored and correlated numerically. DSP has the potential to measure under dynamic testing conditions, because a single recording at each load state is sufficient for the evaluation. Furthermore, the requirements for vibration isolation are much lower than for interferometric methods, because DSP works without reference wave. DSP is therefore an attractive tool for measurements under workshop conditions.

The sample under investigation is coherently illuminated by means of an expanded laser beam. A speckle pattern of the reference state and a speckle pattern of the load state are recorded. The first step of the numerical evaluation procedure is to divide the whole image of e.g. 2,024 × 2,024 pixels into "subimages", Fig. 8.9. The usual sizes of these subimages are 64 × 64 or 32 × 32 pixels. The calculation of the local displacement vectors at each subimage is performed by a cross correlation function

$$R_{II}(d_x, d_y) = \int\limits_{-\infty}^{\infty} \int\limits_{-\infty}^{\infty} I_1^*(x, y) I_2(x + d_x, y + d_y) dx dy \qquad (8.17)$$

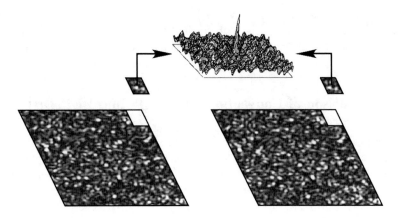

Fig. 8.9 Cross correlation of subimages (from [86])

where $I_1(x, y)$ and $I_2(x, y)$ are the intensities in the reference and in the load speckle pattern, respectively. The quantities d_x and d_y are the displacements of the subimage in x- and y-direction. Intensities are always real i.e. the conjugate complex operation can be neglected. A mathematically equivalent form of Eq. (8.17) is:

$$R_{II}(d_x, d_y) = \Im^{-1}\left\{\Im\left[I_1^*(x, y)\right]\Im\left[I_2(x, y)\right]\right\} \tag{8.18}$$

The mean displacement vector of the evaluated subimage is given by the location of the peak of the cross correlation function, Fig. 8.9. This numerical evaluation corresponds to the classical technique, where double exposed speckle photographs are locally illuminated by a collimated laser beam. The full in-plane displacement map of the monitored area is available after evaluation of all subimages.

The displacement field is calculated by this method in integer numbers of one pixel. The accuracy is therefore only of the order of one pixel. This discrete evaluation is sufficient for applications where only displacements fields are to be measured. Strain analyses of experimental mechanics often require a higher measurement accuracy, to enable the differences to be calculated. The normal strains are e.g. given by

$$\varepsilon_x = \frac{\partial d_x}{\partial x} \approx \frac{\Delta d_x}{\Delta x}; \; \varepsilon_y = \frac{\partial d_y}{\partial y} \approx \frac{\Delta d_y}{\Delta y} \tag{8.19}$$

The accuracy of DSP can be improved using so called subpixel algorithms, where the displacements are calculated on the basis of all floating point values in the neighborhood of the pixel with the peak location. A simple subpixel algorithm is given by

$$d_x = \frac{\sum_i d_{x,i} G_i}{\sum_i G_i} \; d_y = \frac{\sum_i d_{y,i} G_i}{\sum_i G_i} \tag{8.20}$$

where G_i is the floating point grey level of pixel number i. The structure of this formula is equivalent to a "center of gravity" calculation. In practice only a few pixels around the peak are necessary for subpixel evaluation.

8.4 Comparison of Conventional HI, ESPI and Digital HI

Conventional Holographic Interferometry using photographic or other recording media, Electronic Speckle Pattern Interferometry and Digital Holographic Interferometry are all different methods to measure optical path changes. In this section the differences as well as the common features of all three methods are analyzed.

The process flow diagram of conventional real-time HI is shown in Fig. 8.10. The measurement process starts by recording a hologram of the object in its initial

Fig. 8.10 Process flow of real-time HI with phase shifting fringe evaluation

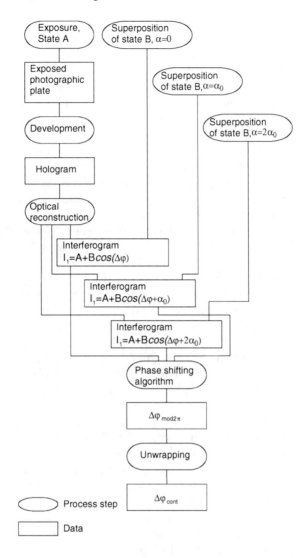

state. This is the most time-consuming and costly step of the entire process: The exposure, development and repositioning of a photographic plate takes typically some minutes. Other holographic recording media such as thermoplastic film or photorefractive crystals can be developed much faster and automated (some seconds for thermoplastic films, even instantaneously for photorefractive crystals). However, the quality and reliability of thermoplastic films is not sufficient for HI applications and the information stored in photorefractive crystals erases in the optical reconstruction process. Once the hologram is successfully developed and replaced at its initial position the process is simple: the superposition of the wave field reconstructed by the hologram with the actual wave field generates a holographic interferogram. This fringe pattern is recorded by an electronic camera and digitally stored. In order to determine the interference phase unambiguously, it is necessary to generate at least three interferograms (for the same object state) with mutual phase shifts. The interference phase is calculated from these phase shifted interferograms by the algorithm briefly discussed in Sect. 2.7.5. The entire process requires altogether the generation of one hologram plus recording of at least three interferograms in order to determine the interference phase. The technical effort is demanding: a holographic set-up with interferometer and laser, holographic recording media (photographic plates), laboratory equipment for development of holograms, a phase shifting unit and an electronic camera with storage device (PC) for interferogram recording are all necessary. On the other hand the quality of interferograms generated by this method is excellent. Due to the size and resolution of holograms recorded on photographic plates the observer can choose the observation direction freely, i.e. it is possible to observe the object from a variety of different positions and with different depth of focus. This is often very helpful for NDT applications and, if the sensitivity vector has to be varied, in quantitative deformation measurement.

ESPI was born from the desire to replace photographic hologram recording and processing by recording with electronic cameras. At the beginning of the 1970s, when ESPI was invented, only analogue cameras with very low resolution (line-pairs per millimetre) were available. Consequently, a direct conversion of holographic principles to electronic recording devices was not possible. The basic idea of ESPI therefore was to record holograms of focussed images. The spatial frequencies of these image plane holograms could be adapted to the resolution of the cameras due to the in-line configuration. The optical reconstruction was replaced by an image correlation (subtraction). The ESPI correlation patterns have some similarities to the fringes of HI, but have a "speckly" appearance. Another difference in comparison to conventional HI is the loss of the 3D-effect, because only image plane holograms are recorded from one observation direction. Interference phase measurement with ESPI requires application of phase shifting methods; see flow process in Fig. 8.11. In each state at least three speckle interferograms with mutual phase shifts have to be recorded. The total number of electronic recordings to determine the interference phase is therefore at least six. Speckle interferometers are commercially available and are nearly as simple to use as ordinary cameras.

The idea of Digital Holographic Interferometry was to record "real" holograms (as opposed to focussed image holograms) by an electronic device and to transfer the optical reconstruction process into the computer. The method is characterized by following features:

- No wet-chemical or other processing of holograms (as for ESPI)
- Different object planes can be reconstructed by numerical methods (numerical focussing) all from one digital hologram
- Lensless imaging is possible, i.e. no aberrations are introduced by imaging devices
- Direct phase reconstruction, i.e. phase differences can be calculated directly from holograms, without interferogram generation and processing. This

Fig. 8.11 Process flow of phase shifting ESPI

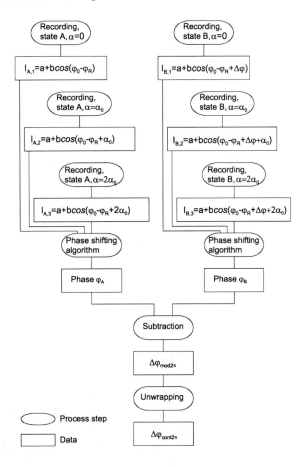

interesting feature is only possible in DHI, conventional HI as well as ESPI need phase shifted interferograms (or another additional information) for phase determination. The total number of recordings to get the interference phase is therefore only two (one per state), see process flow in Fig. 8.12. Even transient processes, where there is no time for recording of phase shifted interferograms, can be investigated with DHI.

DHI and phase shifting ESPI are competing techniques. ESPI allows real-time operation, i.e. the recording speed is only limited by the frame rate of the recording device (CCD). In addition the user directly observes an image of the object under investigation, whereas in DHI this image is only available after running the reconstruction algorithm. This *what you see is what you get* feature is helpful for adjustment and control purposes. On the other hand the time for running DHI reconstruction algorithms has been reduced drastically in recent years due to the progress in computer technology. Digital holograms with 2,000 × 2,000 pixels can nowadays be reconstructed nearly in real-time.

Fig. 8.12 Process flow of
Digital HI

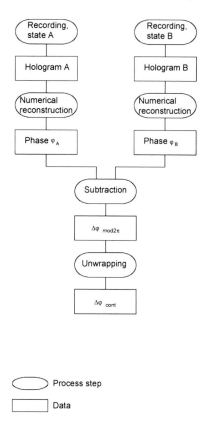

Another *current* disadvantage of DHI is that the spatial frequency spectrum has
to be adapted carefully to the resolution (pixel size) of the sensor. However, the
performance chararctersitics and operational parameters of electronic sensors like
CCD and CMOS are rapidly improving, with pixel sizes for CMOS currently
around 2 μm, which in turn allows recording of larger object cone angles.

Appendix A
The Fourier Transform

A.1 Definitions

The one-dimensional *Fourier transform* of the function $f(x)$ is defined as

$$\Im\{f(x)\} = F(u) = \int\limits_{-\infty}^{\infty} f(x)\exp[-i2\pi ux]dx \qquad (A.1)$$

The inverse one-dimensional *Fourier transformation* is defined as

$$\Im^{-1}\{F(u)\} = f(x) = \int\limits_{-\infty}^{\infty} F(u)\exp[i2\pi ux]du \qquad (A.2)$$

The functions $f(x)$ and $F(u)$ are called *Fourier transform pair*.

The two-dimensional *Fourier transform* of the function $f(x,y)$ is defined as

$$\Im\{f(x,y)\} = F(u,v) = \int\limits_{-\infty}^{\infty}\int\limits_{-\infty}^{\infty} f(x,y)\exp[-i2\pi(ux+vy)]dxdy \qquad (A.3)$$

The corresponding inverse two-dimensional *Fourier transformation* is defined as

$$\Im^{-1}\{F(u,v)\} = f(x,y) = \int\limits_{-\infty}^{\infty}\int\limits_{-\infty}^{\infty} F(u,v)\exp[i2\pi(ux+vy)]dudv \qquad (A.4)$$

The Fourier transformation is a powerful mathematical tool to describe and analyse periodic structures. If x is the time coordinate of a signal (unit: s), then u is the corresponding frequency (unit: $1/s \equiv$ Hz). In the two-dimensional case (x,y) are often spatial coordinates (units: m), while (u,v) are the corresponding spatial frequencies (units: $1/m$).

© Springer-Verlag Berlin Heidelberg 2015
U. Schnars et al., *Digital Holography and Wavefront Sensing*,
DOI 10.1007/978-3-662-44693-5

A.2 Properties

In the following some useful theorems relating to Fourier transforms are summarized. These formulas are given for the two-dimensional case.

1. Linearity theorem

$$\Im\{af(x,y) + bg(x,y)\} = aF(u,v) + bG(u,v) \tag{A.5}$$

where a and b are constants, $F(u,v) = \Im(f(x,y))$ and $G(u,v) = \Im(g(x,y))$.

2. Similarity theorem

$$\Im\{f(ax,by)\} = \frac{1}{|ab|}F\left(\frac{u}{a},\frac{v}{b}\right) \tag{A.6}$$

3. Shift theorem

$$\Im\{f(x-a,y-b)\} = F(u,v)\exp[-i2\pi(ua+vb)] \tag{A.7}$$

4. Rayleigh's (Parseval's) theorem

$$\int\limits_{-\infty}^{\infty}\int\limits_{-\infty}^{\infty}|f(x,y)|^2dxdy = \int\limits_{-\infty}^{\infty}\int\limits_{-\infty}^{\infty}|F(u,v)|^2dudv \tag{A.8}$$

5. Convolution theorem

The two-dimensional convolution of two functions $f(x,y)$ and $g(x,y)$ is defined as

$$(f \otimes g)(x,y) = \int\limits_{-\infty}^{\infty}\int\limits_{-\infty}^{\infty}f(x',y')g(x-x',y-y')dx'dy' \tag{A.9}$$

where $\otimes$ denotes the convolution operation. The convolution theorem states that the Fourier transform of the convolution of two functions is equal to the product of the Fourier transforms of the individual functions:

$$\Im\{f(x,y) \otimes g(x,y)\} = F(u,v)G(u,v) \tag{A.10}$$

6. Autocorrelation theorem

$$\Im\left\{\int\limits_{-\infty}^{\infty}\int\limits_{-\infty}^{\infty}f^*(x',y')f(x+x',y+y')dx'dy'\right\} = |F(u,v)|^2 \tag{A.11}$$

7. Fourier integral theorem

$$\Im\Im^{-1}\{f(x,y)\} = \Im^{-1}\Im\{f(x,y)\} = f(x,y) \tag{A.12}$$

8. Differentiation

Differentiation in the spatial domain corresponds to a multiplication with a linear factor in the spatial frequency domain:

$$\Im\left\{\left(\frac{\partial}{\partial x}\right)^m \left(\frac{\partial}{\partial y}\right)^n f(x,y)\right\} = (i2\pi u)^m (i2\pi v)^n F(u,v) \tag{A.13}$$

A.3 The Discrete Fourier Transform

For numerical computations the function to be transformed is given in a discrete form, i.e. $f(x)$ in Eq. (A.1) has to be replaced by the finite series f_k, with integer numbers $k = 0, 1, \ldots, N - 1$. The continuous variable x is now described as integer multiple of a sampling interval Δx:

$$x = k\Delta x \tag{A.14}$$

The frequency variable u is converted into a discrete variable, too:

$$u = m\Delta u \tag{A.15}$$

The discrete representation of Eq. (A.1) is then given by:

$$F_m = \Delta x \sum_{k=0}^{N-1} f_k \exp[-i2\pi km\Delta x\Delta u] \quad \text{for } m = 0, 1, \ldots, N-1 \tag{A.16}$$

The maximum frequency is determined by the sampling interval in the spatial domain:

$$u_{\max} = N\Delta u = \frac{1}{\Delta x} \tag{A.17}$$

The following expression

$$F_m = \frac{1}{N}\sum_{k=0}^{N-1} f_k \exp\left[-i2\pi\frac{km}{N}\right] \tag{A.18}$$

is therefore defined as one-dimensional *discrete Fourier transform* (DFT). The inverse transformation is given by

$$f_k = \sum_{m=0}^{N-1} F_m \exp\left[i2\pi\frac{km}{N}\right] \tag{A.19}$$

Similar considerations lead to the discrete two-dimensional Fourier transform pair:

$$F_{mn} = \frac{1}{N^2} \sum_{k=0}^{N-1} \sum_{l=0}^{N-1} f_{kl} \exp\left[-i2\pi\left(\frac{km+ln}{N}\right)\right] \tag{A.20}$$

$$f_{kl} = \sum_{m=0}^{N-1} \sum_{n=0}^{N-1} F_{mn} \exp\left[i2\pi\left(\frac{km+ln}{N}\right)\right] \tag{A.21}$$

for $m = 0, 1, \ldots N - 1$ and $n = 0, 1, \ldots N - 1$

Here a quadratic field of sampling points is used, i.e. the number of points in each row is equal to that in each column.

The computation time for a discrete Fourier transform is mainly determined by the number of complex multiplications. A two-dimensional DFT can be factorised into two one-dimensional DFT's:

$$F_{mn} = \frac{1}{N^2} \sum_{k=0}^{N-1} \left[\sum_{l=0}^{N-1} f_{kl} \exp\left(-i2\pi\frac{nl}{N}\right) \right] \exp\left(-i2\pi\frac{km}{N}\right) \tag{A.22}$$

The one-dimensional DFT can be programmed most effectively using the so called *fast fourier transform* (FFT) algorithms, invented in the 70th of the last century by Cooley and Tookey. These algorithms make use of redundancies and reduce the number of multiplications for a one-dimensional DFT from N^2 to $2N \log_2 N$. The FFT algorithms are not described here, it is referred to the literature [18].

Appendix B
Phase Transformation of a Spherical Lens

B.1 Lens Transmission Function

The effect of an optical component with refractive index n and thickness d on the complex amplitude of a wave is described by a transmission function τ:

$$\tau = |\tau| \exp\left[-i\frac{2\pi}{\lambda}(n-1)d\right] \tag{B.1}$$

This function is calculated in the following for a thin biconvex lens. Such lens consists of two spherical surfaces, see Fig. B.1. The radius of curvature of the left half lens is R_1, while that of the right half lens is designated R_2. Following sign convention is applied: As rays travel from left to right, each convex surface has a positive radius of curvature, while each concave surface has a negative radius of curvature. Due to this convention R_2 has a negative value. Losses due to reflection at the surfaces and due to absorption inside the lens are neglected; i.e. $|\tau| = 1$. The refractive index is constant for the entire lens.

The lens thickness is a function of the spatial coordinates x and y:

$$\begin{aligned} d(x,y) &= d_1(x,y) + d_2(x,y) \\ &= d_{01} - \zeta_1 + (d_{02} - \zeta_2) \end{aligned} \tag{B.2}$$

According to Fig. B.1 it can be written:

$$\begin{aligned} R_1^2 &= r^2 + (R_1 - \zeta_1)^2 \\ &= x^2 + y^2 + R_1^2 - 2R_1\zeta_1 + \zeta_1^2 \end{aligned} \tag{B.3}$$

and

$$\begin{aligned} R_2^2 &= r^2 + (-R_2 - \zeta_2)^2 \\ &= x^2 + y^2 + R_2^2 + 2R_2\zeta_2 + \zeta_2^2 \end{aligned} \tag{B.4}$$

© Springer-Verlag Berlin Heidelberg 2015
U. Schnars et al., *Digital Holography and Wavefront Sensing*,
DOI 10.1007/978-3-662-44693-5

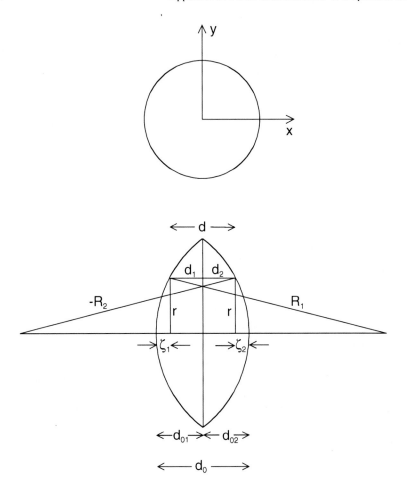

Fig. B.1 Biconvex lens, *top* view and cross-sectional view

Neglecting the quadratic terms of $\zeta_{1/2}$ leads to:

$$\zeta_1 = \frac{x^2 + y^2}{2R_1} \tag{B.5}$$

$$\zeta_2 = -\frac{x^2 + y^2}{2R_2} \tag{B.6}$$

This level of approximation is consistent with the parabolic approximation used in the Fresnel transformation. The thickness is now

$$d(x, y) = d_0 - \frac{x^2 + y^2}{2R_1} + \frac{x^2 + y^2}{2R_2} \tag{B.7}$$

With the lens makers equation

$$\frac{1}{f} = (n - 1)\left[\frac{1}{R_1} - \frac{1}{R_2}\right] \tag{B.8}$$

of geometrical optics following lens transmission function is derived:

$$L(x, y) = \exp\left[i\frac{\pi}{\lambda f}(x^2 + y^2)\right] \tag{B.9}$$

The constant factor $\exp(-i2\pi/\lambda(n - 1)d_0)$, which only effects the overall phase, has been neglected.

B.2 Correction of Aberrations

In the following the complex amplitude of an object, which is imaged by a lens is calculated. The object is lying in the (ξ, η) coordinate system, the lens is located in the (x, y) system and the image arises in the (ξ', η') system, see Fig. B.2. The object is described by the complex amplitude $E_O(\xi, \eta)$.

The complex amplitude in front of the lens is given by

$$E'_O(x, y) = \exp\left[-i\frac{\pi}{\lambda d}(x^2 + y^2)\right] \int_{-\infty}^{\infty} \int_{-\infty}^{\infty} E_O(\xi, \eta) \exp\left[-i\frac{\pi}{\lambda d}(\xi^2 + \eta^2)\right]$$

$$\times \exp\left[i\frac{2\pi}{\lambda d}(x\xi + y\eta)\right] d\xi d\eta \tag{B.10}$$

where the Fresnel approximation is used. The complex amplitude in the image plane in then given by

$$E''_O(\xi', \eta') = \exp\left[-i\frac{\pi}{\lambda d}(\xi'^2 + \eta'^2)\right] \int_{-\infty}^{\infty} \int_{-\infty}^{\infty} E'_O(x, y)L(x, y) \exp\left[-i\frac{\pi}{\lambda d}(x^2 + y^2)\right]$$

$$\times \exp\left[i\frac{2\pi}{\lambda d}(x\xi' + y\eta')\right] dx dy$$

$$= \exp\left[-i\frac{\pi}{\lambda d}(\xi'^2 + \eta'^2)\right] \int_{-\infty}^{\infty} \int_{-\infty}^{\infty} \int_{-\infty}^{\infty} \int_{-\infty}^{\infty} E_O(\xi, \eta) \exp\left[-i\frac{\pi}{\lambda d}(x^2 + y^2)\right]$$

$$\times \exp\left[-i\frac{\pi}{\lambda d}(\xi^2 + \eta^2)\right] \exp\left[i\frac{2\pi}{\lambda d}(x\xi + y\eta)\right] \exp\left[i\frac{2\pi}{\lambda d}(x^2 + y^2)\right]$$

$$\times \exp\left[-i\frac{\pi}{\lambda d}(x^2 + y^2)\right] \exp\left[i\frac{2\pi}{\lambda d}(x\xi' + y\eta')\right] d\xi d\eta dx dy \tag{B.11}$$

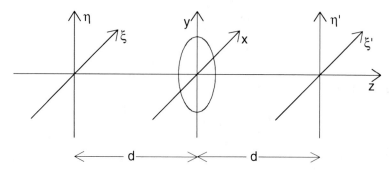

Fig. B.2 Image formation

A magnification of 1 and a focal distance of $f = d/2$ is used for the lens transmission function $L(x,y)$.

The image coordinates can be expressed in terms of the object coordinates:

$$\xi' = -\xi \quad \text{and} \quad \eta' = -\eta \tag{B.12}$$

The minus signs result, because according to the laws of geometrical optics the image is upside down.

The complex amplitude of the image is now

$$E_O''(\xi', \eta') = \exp\left[-i\frac{2\pi}{\lambda d}\left(\xi'^2 + \eta'^2\right)\right] E_O(-\xi', -\eta')$$

$$= \exp\left[-i\frac{\pi}{\lambda f}\left(\xi'^2 + \eta'^2\right)\right] E_O(-\xi', -\eta') \tag{B.13}$$

The wavefield in the image plane has to be multiplied therefore by a factor

$$P(\xi', \eta') = \exp\left[i\frac{\pi}{\lambda f}\left(\xi'^2 + \eta'^2\right)\right] \tag{B.14}$$

in order to generate the correct phase distribution.

This correction factor depends on the wavelength and on the coordinates of the image plane. It can be neglected, if only the intensity of a wavefield has to be calculated after reconstruction ($I \propto P^*P = 1$). This is also valid if the phase difference of two wavefields, which are recorded with the *same* wavelength, is computed:

$$\Delta\varphi = \varphi_1 - \varphi_2 = i\pi/\lambda f\left(\xi'^2 + \eta'^2\right) + \varphi_1' - \left[i\pi/\lambda f\left(\xi'^2 + \eta'^2\right) + \varphi_2'\right]$$

$$= \varphi_1' - \varphi_2' \tag{B.15}$$

This is usually the case in DHI for applications in deformation analysis. However, the correction factor has to be considered, if the phase difference of two wavefields, which are recorded with *different* wavelengths, is computed. This is the case in multi-wavelength DHI for shape measurement.

Appendix C
Simple Reconstruction Routines

Two simple Matlab© reconstruction routines are given here. They may easily be converted into other programming languages. In Fig. C.1 the Fresnel transform is shown, reconstruction according to the convolution approach is depicted in Fig. C.2.

© Springer-Verlag Berlin Heidelberg 2015
U. Schnars et al., *Digital Holography and Wavefront Sensing*,
DOI 10.1007/978-3-662-44693-5

```
%Hologram rekonstruction, Fresnel transformation
I=imread(,hologram.BMP');
imshow(I);
C=complex(I);
C=double(C);
lam=632.817e-9;
%lam=wavelength
d=1.054;
%d=reconstruction distance
dx=6.8e-6;
%dx=pixel size
for k=1:1024
    for l=1:1024
        C(k,l)=C(k,l)*exp(-i*pi/lam/d*(k*k*dx*dx+l*l*dx*dx));
    end;
end;
D=ifft2(C);
%inverse FFT
F=abs(D);
% re-sort of image quadrants
for k=1:512
kk=k+512;
for l=1:512
ll=l+512;
x1=F(k,l);
x2=F(kk,ll);
F(k,l)=x2;
F(kk,ll)=x1;
x3=F(kk,l);
x4=F(k,ll);
F(kk,l)=x4;
F(k,ll)=x3;
end;
end;
imshow(F)
```

Fig. C.1 Reconstruction routine, Fresnel Transformation

```
%Hologram reconstruction, convolution approach
I=imread(,hologram.BMP');
imshow(I);
C=complex(I);
C=double(C);
lam=632.817e-9;
%lam=wavelength
d=1.054;
%d=recording distance
mag=1/7;
% mag=magnification
d2=d*mag;
% d2=reconstruction distance
f=1/(1/d+1/d2);
% f=focal distance of numerical lens
dx=6.8e-6;
%dx=pixel size
% calculation of product between hologram function and
% lens transmission function
for k=1:1024
    for l=1:1024
        C(k,l)=C(k,l)*exp(i*pi/lam/f*(k*k*dx*dx+l*l*dx*dx));
    end;
end;
D=fft2(C);
% FFT
for k=1:1024
    for l=1:1024
        g(k,l)=i/lam*exp(-i*pi*2/lam*sqrt(d2*d2+(k-
512)^2*dx*dx+(l-512)^2*dx*dx))/sqrt(d2*d2+(k-512)^2*dx*dx+(l-
512)^2*dx*dx);

    end;
end;
G=fft2(g);
C=ifft2(D.*G);
H=log(abs(C));

imagesc(H)
```

Fig. C.2 Reconstruction routine, convolution approach

References

1. Abramson N (1983) Light-in-flight recording: high speed holographic motion pictures of ultrafast phenomena. Appl Opt 22:215–232
2. Abramson N (1984) Light-in-flight recording 2: compensation for the limited speed of the light used for observation. Appl Opt 23:1481–1492
3. Abramson N (1984) Light-in-flight recording 3: compensation for optical relativistic effects. Appl Opt 23:4007–4014
4. Abramson N (1985) Light-in-flight recording 4: visualizing optical relativistic phenomena. Appl Opt 24:3323–3329
5. Adams M, Kreis T, Jüptner W (1997) Particle size and position measurement with digital holography. Proc SPIE 3098:234–240
6. Adams M, Kreis T, Jüptner W (1999) Particle measurement with digital holography. In: Proc SPIE vol 3823 38-43
7. Agour M, Falldorf C, Bergmann R (2013) Investigation of composite materials using SLM-based phase retrieval. Opt Lett 38:2203–2205
8. Allen LJ, Oxley MP (2001) Phase retrieval from series of images obtained by defocus variation. Opt Comm 199:65–75
9. Almoro PF, Maallo A, Hanson SG (2009) Fast convergent algorithm for speckle-based phase retrieval and a design for dynamic wavefront sensing. Appl Opt 48:1485–1493
10. Arroyo K, Hinsch (2008) Recent developments of PIV towards 3D measurements. In: Schröder A, Willert CE (eds.) Particle image velocimetry. Springer, Berlin, pp. 127–154
11. Baranova NB, Mamaev AV, Pilipetsky NF, Shkunov VV, Zel'dovich BY (1983) Wave-front dislocations: topological limitations for adaptive systems with phase conjugation". J Opt Soc Am 73:525–528
12. Bauschke HH, Combettes PL, Luke DR (2002) Phase retrieval, error reduction algorithm, and Fienup variants: a view from convex optimization. J Opt Soc Am A 19:1334–1345
13. Bauschke HH, Combettes PL, Luke DR (2003) Hybrid projection-reflection method for phase retrieval. J Opt Soc Am A 20:1025–1034
14. Le Bigot E-O, Wild WJ, Kibblewhite EJ (1998) Reconstruction of discontinuous light-phase functions. Opt Lett 23:10–12
15. Bisle W (1998) Optische prüfung an luftfahrtkomponenten: weiterentwicklung des scherografie-prüfverfahrens für nicht-kooperative oberflächen von flugzeugstrukturen. Proc Deutsche Gesellschaft für Luft- und Raumfahrt, annual meeting, Bremen, Germany
16. Born M, Wolf E (1980) Principles of Optics. 6 edn. Oxford, Pergamon
17. Breuckmann B, Thieme W (1985) Computer-aided analysis of holographic interferograms using the phase-shift method. Appl Opt 24:2145–2149
18. Brigham EO (1974) The fast fourier transform. Pretince-Hall, New York

19. van Brug H (1997) Zernike polynomials as a basis for wave-front fitting in lateral shearing interferometry. Appl. Opt. 36:2788–2790
20. Bryngdahl O, Wyrowski F (1990) Digital Holography—computer generated holograms. Progress in Optics 28:1–86
21. Burns NJ, Watson J (2011) A study of focus metrics and their application to automated focusing of inline transmission holograms. Jnl Imaging Science 59:90–99
22. Burns NM (2011) Automated analysis system for the study of digital in-line holograms of aquatic particles, PhD Thesis, Univ of Aberdeen
23. Butters JN, Leendertz JA (1971) Holographic and Videotechniques applied to engineering measurements. J Meas Control 4:349–354
24. Carder K (1979) Holographic microvelocimeter for use in studying ocean particle dynamics. Opt Eng 18:524–525
25. Carlsson T, Nilsson B, Gustafsson J (2001) System for acquisition of three-dimensional shape and movement using digital Light-in-Flight holography. Opt Eng 40(01):67–75
26. CCD Primer (2002) product information. Kodak, New York
27. Champagne EB (1967) Non-paraxial imaging, magnification and aberration properties in holography. J Op Soc Am 57:51–55
28. Champeney DC (1973) Fourier transforms and their physical interpretation. Academic Press, London
29. Chen DJ, Chiang FP, Tan YS, Don HS (1993) Digital speckle displacement measurement using a complex spectrum method. Appl Opt 32(11):1839–1848
30. Claus D, Watson J, Rodenburg J (2011) Analysis and interpretation of the Seidel aberration coefficients in digital holography. App Opt 50:H220–H229
31. Coppola G, De Nicola S, Ferraro P, Finizio A, Grilli S, Iodice M, Magro C, Pierattini G (2003) Evaluation of residual stress in MEMS structures by means of digital holography. In: Proc. SPIE vol 4933, pp 226-31
32. Colomb T, Kühn J, Charriére F, Depeursinge C, Marquet P, Aspert N (2006) Optics Express 14:4300–4306
33. Coppola G, Ferraro P, Iodice M, De Nicola S, Finizio A, Grilli S (2004) A digital holographic microscope for complete characterization of microelectromechanical systems. Meas Sci Technol 15:529–539
34. Coquoz O, Conde R, Taleblou F, Depeursinge C (1995) Performances of endoscopic holography with a multicore optical fiber. Appl Opt 34(31):7186–7193
35. Creath K (1985) Phase shifting speckle-interferometry. Appl Opt 24(18):3053–3058
36. Creath K (1994) Phase-shifting holographic interferometry. Holographic Interferometry, Springer Series in Optical Sciences 68:109–150
37. Cuche E, Bevilacqua F, Depeursinge C (1999) Digital holography for quantitative phase-contrast imaging. Optics Letters 24(5):291–293
38. Cuche E, Marquet P, Depeursinge C (1999) Simultaneous amplitude-contrast and quantitative phase-contrast microscopy by numerical reconstruction of Fresnel off-axis holograms. Appl Opt 38(34):6994–7001
39. Cuche E, Marquet P, Depeursinge C (2000) Spatial filtering for zero-order and twin-image elimination in digital off-axis holography. Appl Opt 39(23):4070–4075
40. Cuche E, Marquet P, Depeursinge C (2000) Aperture apodization using cubic spline interpolation: application in digital holographic microscopy. Opt Commun 182:59–69
41. Demetrakopoulos TH, Mittra R (1974) Digital and optical reconstruction of images from suboptical diffraction patterns. Appl Opt 13(3):665–670
42. Demoli N, Mestrovic J, Sovic I (2003) Subtraction digital holography. Appl Opt 42 (5):798–804
43. Depeursinge C, Marquet P, Pavillon P (2011) Applications of digital holographic microscopy in biomedicine, in Handbook of Biomedical Optics, DA Boas, P Pitros, N Ramanujam (eds), Taylor and Francis, 29: 617-647

44. Dong H, Khong C, Player MA, Solan M, Watson J (2003) Algorithms and applications for electronically-recorded holography. Proc SPIE 5477:354–365
45. Dong BZ, Zhang Y, Gu BY, Yang GZ (1997) Numerical investigation of phase retrieval in a fractional Fourier transform. J. Opt. Soc. Am. A 14:2709–2714
46. Doval AF (2000) A systematic approach to TV holography. Meas Sci Technol 11:R1–R36
47. Dubois F, Joannes L, Legros JC (1999) Improved three-dimensional imaging with a digital holography microscope with a source of partial spatial coherence. Appl Opt 38 (34):7085–7094
48. Dubois F, Minetti C, Monnom O, Yourassowsky C, Legros JC, Kischel P (2002) Pattern recognition with a digital holographic microscope working in partially coherent illumination. Appl Opt 41(20):4108–4119
49. Dubois F, Schockaert C, Callens N Yourassowsky C (2006) Focus plane detection criteria in digital holography microscopy by amplitude analysis. Opt Express 14: 5895–5908
50. Elser V (2003) Phase retrieval by iterated projections. J. Opt. Soc. Am A 20:40–55
51. Elster C, Weingärtner I (1999) Solution to the Shearing Problem. Appl. Opt. 38:5024–5031
52. Falldorf C, Huferath-von Luepke S, von Kopylow C, Bergmann R (2012) Reduction of speckle noise in multiwavelength contouring. Appl. Opt. 51(34):8211–8215
53. Falldorf C, Agour M, von Kopylow C, Bergmann R (2010) Phase retrieval by means of a spatial light modulator in the Fourier domain of an imaging system. Appl. Opt. 49:1826–1830
54. Falldorf C, Heimbach Y, von Kopylow C, Jüptner W (2007) Efficient reconstruction of spatially limited phase distributions from their sheared representation. Appl. Opt. 46:5038–5043
55. Falldorf C, Osten W, von Kopylow C, Jüptner W (2009) Shearing interferometer based on the birefringent properties of a spatial light modulator. Opt. Lett. 34:2727–2729
56. Falldorf C, Simic A, Ehret G, Schulz M, von Kopylow C, Bergmann R et al (2014) Precise optical metrology using computational shear interferometry and an LCD monitor as light source. Fringe 2013, 7th International Workshop on Advanced Opt Imaging Metrol 729–734
57. Falldorf C, von Kopylow C, Bergmann R (2013) Wave field sensing by means of computational shear interferometry. J. Opt. Soc. Am. A 30:1905–1912
58. Fienup J, Wackerman C (1986) Phase-retrieval stagnation problems and solutions. J Opt Soc Am A 3:1897–1907
59. Fienup J (1982) Phase retrieval algorithms: a comparison. Appl Opt 21:2758–2769
60. Fienup J (1993) Phase-retrieval algorithms for a complicated optical system. Appl Opt 32:1737–1746
61. Fienup J (1978) Reconstruction of an object from the modulus of its Fourier transform. Opt. Lett. 3:27–29
62. Frank J, Altmeyer S, Wernicke G (2010) Non-interferometric, non-iterative phase retrieval by Green's functions. J Opt Soc Am A 70:2244–2251
63. Frauel Y, Javidi B (2001) Neural network for three-dimensional object recognition based on digital holography. Optics Letters 26(19):1478–1480
64. Frauel Y, Tajahuerce E, Castro MA, Javidi B (2001) Distortion- tolerant three-dimensional object recognition with digital holography. Appl Opt 40(23):3887
65. Freischlad KR, Koliopoulos CL (1986) Modal estimation of a wave front from difference measurements using the discrete Fourier transform. J Opt Soc Am A 3:1852–1861
66. Fried DL (2001) Adaptive optics wave function reconstruction and phase unwrapping when branch points are present. Opt Commun 200:43–72
67. Füzessy Z, Gyimesi F (1984) Difference holographic interferometry: displacement measurement. Opt Eng 23(6):780–783
68. Gabor D (1948) A new microscopic principle. Nature 161:777–778
69. Gabor D (1949) Microscopy by reconstructed wavefronts. Proc Roy Soc 197:454–487
70. Gabor D (1951) Microscopy by reconstructed wavefronts: 2. Proc Phys Soc 64:449–469

71. Gerchberg RW, Saxton WO (1972) A practical algorithm for the determination of phase from image and diffraction plane pictures. Optik (Jena) 35:237–246

72. Goodman JW (1975) statistical properties of laser speckle patters. In: Dainty JC (ed) Laser Speckle and Related Phenomena, Topics in Appl Physics, vol 9. Springer, Berlin, pp 9–75

73. Goodman JW (1996) Introduction to fourier optics 2nd edn. McGraw-Hill, New York

74. Goodman JW, Lawrence RW (1967) Digital image formation from electronically detected holograms. Appl Phys Lett 11:77–79

75. Grilli S, Ferraro P, De Nicola S, Finizio A, Pierattini G, Meucci R (2001) Whole optical wavefield reconstruction by digital holography. Optics Express 9(6):294–302

76. Grunwald R, Huferath S, Bock M, Neumann U, Langer S (2007) Angular tolerance of Shack-Hartmann wavefront sensors with microaxicons. Opt. Lett. 32:1533–1535

77. Gureyev T, Nugent KA (1996) Phase retrieval with the transport-of-intensity equation. II. Orthogonal series solution for nonuniform illumination. J. Opt. Soc. Am. A 13:1670–1682

78. Haddad W, Cullen D, Solem J, Longworth J, McPherson A, Boyer K, Rhodes K (1992) Fourier-transform holographic microscope. Appl Opt 31(24):4973–4978

79. Harriharan P (1984) Optical Holography. Cambridge University Press, Cambridge

80. Hartmann J (1900) Bemerkungen über den Bau und die Justierung von Spektrographen. Zeitschrift für Instrumentenkunde 20: 47ff

81. Heath JP (2005) Dictionary of microscopy Wiley, UK

82. Helmers H, Bischoff M, Ehlkes L (2001) ESPI-system with active in-line digital phase stabilization. In: Jüptner W, Osten W (eds) Proc 4th international workshop on automatic processing of fringe patterns. Elsevier, Heidelberg, pp 673–679

83. Hinsch K (2002) Holographic particle image velocimetry. Meas Sci Technol 13:R61–R72

84. Hinsch KD, Herrmann SV(eds) (2004) Special Issue on "Holographic particle image velocimetry. in measurement science and technology 15(4): 673–685

85. Hobson PR, Watson J (2002) The principles and practice of holographic recording of plankton. Jnl of Optics A Pure and Applied Optics 4:S34–S49

86. Holstein D, Hartmann HJ, Jüptner W (1998) Investigation of Laser Welds by Means of Digital Speckle Photography. Proc SPIE 3478:294–301

87. Hongbin Y, Guangya Z, Siong CF, Feiwen L, Shouhua W (2008) A tunable Shack-Hartmann wavefront sensor based on a liquid-filled microlens array. J. Micromech. Microeng. 18:105017

88. Hudgin RH (1977) Wave-front reconstruction for compensated imaging. J. Opt. Soc. Am. A 67:357–378

89. Hung YY (1996) Shearography for non-destructive evaluation of composite structures. Opt Lasers Eng 24:161–182

90. Hung YY, Liang CY (1979) Image shearing camera for direct measurement of surface strain. Appl Opt 18:1046–1051

91. Huntley JM, Saldner H (1993) Temporal phase-unwrapping algorithm for automated interferogram analysis. Appl Opt 32(17):3047–3052

92. Inomato O, Yamaguchi I (2001) Measurements of Benard-Marangoni waves using phase-shifting digital holography. Proc SPIE 4416:124–127

93. Ivanov VY, Sivokon VP, Vorontsov MA (1992) Phase retrieval from a set of intensity measurements: theory and experiment. J. Opt. Soc. Am. A 9:1515–1524

94. Jacquot M, Sandoz P, Tribillon G (2001) High resolution digital holography. Opt Commun 190:87–94

95. Javidi B, Nomura T (2000) Securing information by use of digital holography. Optics Letters 25(1):28–30

96. Javidi B, Tajahuerce E (2000) Three-dimensional object recognition by use of digital holography. Optics Letters 25(9):610–612

97. Jericho SK, Garcia-Sucerquia J, Xu W, Jericho MH, Kreuzer HJ (2006) Submersible digital in-line holographic microscope. Rev Sci Instrum 77: 043706-1–043706-10

98. Jüptner W (1978) Automatisierte Auswertung holografischer Interferogramme mit dem Zeilen-Scanverfahren. In: Kreitlow H, Jüptner W (eds) Proc Frühjahrsschule 78 Holografische Interferometrie in Technik und Medizin

99. Jüptner W, Kreis T, Kreitlow H (1983) Automatic evaluation of holographic interferograms by reference beam phase shifting. Proc SPIE 398:22–29

100. Jüptner W, Pomarico J, Schnars U (1996) Light-in-Flight measurements by Digital Holography. In: Proc. SPIE vol 2860, pp. 22–29

101. Jüptner W (2002) Digital Holography: Techniques and Sensors in Microsystem Engineering. Proc. SPIE, (Seattle)

102. Jüptner W, Osten W (2002) Coherent shape control using coherent masks. SPIE proceedings, pp. 338–350.

103. Kato J, Yamaguchi I, Matsumura T (2002) Multicolor digital holography with an achromatic phase shifter. Opt Lett 27(16):1403–1405

104. Katz J (1999) Submersible holocamera for detection of particle characteristics and motions in the sea. Deep Sea Res. Instrum. Methods 46:1455–1481

105. Kebbel V, Grubert B, Hartmann HJ, Jüptner W, Schnars U (1998) Application of digital holography to space-borne fluid science measurements. In: Proc 49th International astronautical congress melbourne paper no. IAF-98-J.5.03

106. Kebbel V, Adams M, Hartmann HJ, Jüptner W (1999) Digital holography as a versatile optical diagnostic method for microgravity experiments. Meas Sci Technol 10:893–899

107. Kebbel V, Hartmann HJ, Jüptner W (2001) Application of digital holographic microscopy for inspection of micro-optical components. Proc SPIE 4398:189–198

108. Kemper B, von Langehanenberg P, Bally G (2007) Digital Holographic Microscopy. Optik & Photonik 2:41–44

109. Kilpatrick JM, Watson J (1993) Underwater hologrammetry: reduction of aberrations by index compensation. J Phys D: App Phys 26:177–182

110. Kilpatrick JM, Watson J (1994) Precision replay of underwater holograms. Meas Sci Technol 5:716–725

111. Kim MK (1999) Wavelength-scanning digital interference holography for optical section imaging. Optics Letters 24(23):1693–1695

112. Kim MK (2000) Tomographic three-dimensional imaging of a biological specimen using wavelength-scanning digital interference holography. (2000). Optics Express 7(9):305–310

113. Kim S, Lee B, Kim E (1997) Removal of bias and the conjugate image in incoherent on-axis triangular holography and real-time reconstruction of the complex hologram. Appl Opt 36 (20):4784–4791

114. Kim MK (2010) Applications of Digital Holography in Biomedical Microscopy. Journal of the Optical Society of Korea 14(2):77–89

115. King RA (1989) The use of self-entropy as a focus measure in digital holography. Pattern Recognition Letters 9:19–25

116. Klein MV, Furtak TE (1986) Optics, 2nd edn. Wiley, New York

117. Knox C (1966) Holographic microscopy as a technique for recording dynamic microscopic subjects. Science 153:989–990

118. Knox C, Brooks RE (1969) Holographic motion picture microscopy. Proc Roy Soc B 174:115–121

119. Kolenovic E, Lai S, Osten W, Jüptner W (2001) Endoscopic shape and deformation measurement by means of Digital Holography. In: Jüptner W, Osten W (eds) Proc 4th International Workshop on Automatic Processing of Fringe Patterns. Akademie, Berlin, pp 686–691

120. Kolenovic E (2005) Correlation between intensity and phase in monochromatic light. J. Opt. Soc. Am. A 22:899–906

121. Kreis T (1996) Holographic Interferometry. Akademie, Berlin

122. Kreis T (2002) Frequency analysis of digital holography. Opt Eng 41(4):771–778

123. Kreis T (2002) Frequency analysis of digital holography with reconstruction by convolution. Opt Eng 41(8):1829–1839
124. Kreis T, Jüptner W (1997) Principles of digital holography. In: Jüptner W, Osten W (eds) Proc 3rd International Workshop on Automatic Processing of Fringe Patterns. Akademie, Berlin, pp 353–363
125. Kreis T, Jüptner W (1997) Suppression of the dc term in digital holography. Opt Eng 36 (8):2357–2360
126. Kreis T, Jüptner W, Geldmacher J (1998) Digital Holography: Methods and Applications. Proc SPIE 3407:169–177
127. Kreis T, Adams M, Jüptner W (1999) Digital in-line holography in particle measurement. In: Proc SPIE vol 3744, 54–. 64
128. Kreis T, Aswendt P, Höfling R (2001) Hologram reconstruction using a digital micromirror device. Opt Eng 40(6):926–933
129. Kreis T, Adams M, Jüptner W (2002) Aperture synthesis in digital holography. Proc SPIE 4777:69–76
130. Kreis T (2005) Handbook of holographic interferometry. Wiley VCH, Weinheim
131. Kreuzer HJ, Pawlitzek RA (1997) Numerical Reconstruction for in-line Holography in Reflection and under glancing Incidence. In: Jüptner W, Osten W (eds) Proc 3rd International Workshop on Automatic Processing of Fringe Patterns. Akademie, Berlin, pp 364–367
132. Kronrod MA, Merzlyakov NS, Yaroslavski LP (1972) Reconstruction of holograms with a computer. Sov Phys-Tech Phys USA 17(2):333–334
133. Kujawinska M, Kozacki T, Falldorf C, Meeser T, Henelly BM, Garbat P, Zaperty W, Niemela M, Finke G, Kowiel M, Naughton T (2014) Multiwavefront digital holographic television. Optics Express 22(3):2324–2336
134. Kulak M, Pisarek J (2001) Speckle photography in the examination of composites. In: W Jüptner , W Osten (eds) Proc 4th International workshop on automatic processing of fringe patterns. Elsevier, pp 528–530
135. Lai S, Neifeld M (2000) Digital wavefront reconstruction and its application to image encryption. Opt Commun 178:283–289
136. Lai S, Kemper B, von Bally G (1999) Off-axis reconstruction of in-line holograms for twin-image elimination. Optics Communications 169:37–43
137. Lai S, King B, Neifeld MA (2000) Wave front reconstruction by means of phase-shifting digital in-line holography. Optics Communications 173:155–160
138. Lai S, Kolenovic E, Osten W, Jüptner W (2002) A deformation and 3D-shape measurement system based on phase-shifting digital holography. Proc SPIE 4537:273–276
139. Latta JN (1971) Comupter-based analysis of holography using ray tracing. App Opt 10:2698–2710
140. Lee WH (1978) Computer-generated Holograms: Techniques and Applications. Progress in Optics 16:120–232
141. Leith EN, Upatnieks J (1962) Reconstructed wavefronts and communication theory. J Opt Soc Am 52:1123–1130
142. Leith EN, Upatnieks J (1964) Wavefront reconstruction with diffused illumination and threedimensional objects. J Opt Soc Am 54:1295–1301
143. Levi A, Stark H (1984) Image restoration by the method of generalized projections with application to restoration from magnitude. J. Opt. Soc. Am. A 1:932–943
144. Liebling M, Unser M (2004) Autofocus for digital Fresnel holograms by use of a Fresnelet-sparsity criterion. J Opt Soc Am A 21:2424–2430
145. Li W, Loomis NC, Hu Q, Davis CS (2007) Focus detection from digital in-line holograms based on spectral l-1 norms. Journal of the Optical Society of America A 24:3054–3062
146. Liu G, Scott PD (1987) Phase retrieval and twin-image elimination for in-line Fresnel holograms. J Opt Soc Am A 4(1):159–165
147. Lokberg O (1980) Electron Speckle Pattern Interferometry. Phys Technol 11:16–22

148. Lokberg O, Slettemoen GA (1987) Basic Electronic Speckle Pattern Interferometry. Appl Opt Eng 10:455–505
149. Luke D (2005) Relaxed averaged alternating reflections for diffraction imaging. Inverse Problems 21:37–50
150. Macovski A, Ramsey D, Schaefer LF (1971) Time Lapse Interferometry and Contouring using Television Systems. Appl. Opt. 10(12):2722–2727
151. Maiman T (1960) Stimulated optical radiation in ruby. Nature 187:493
152. Malkiel E, Abras JN, Katz J (2003) Automated scanning and measurements of particle distribution within a holographic reconstructed volume. Meas Sci Technol 15:601–612
153. Malkiel E (2003) The three-dimensional flow field generated by a feeding calnoid copepod measured using digital holography. J Exp Biol 206:3657–3666
154. Matoba O, Naughton TJ, Frauel Y, Bertaux N, Javidi B (2002) Real-time three-dimensional object reconstruction by use of a phase-encoded digital hologram. Appl Opt 41 (29):6187–6192
155. Meier RW (1965) Magnification and third-order theory in holography. J Op Soc Am 55:987–992
156. Meng H, Hussain F (1995) In-line recording and off-axis viewing technique for holographic particle velocimetry. App Opt 34:1827–1840
157. Meng, G. Pan, Y. Pu, SH. Woodward (2004) Holographic particle image velocimetry: from film to digital recording. Meas Sci Technol 15: 673–685
158. Misell DL (1973) A method for the solution of the phase problem in electron Microscopy. J. Phys. D Appl. Phys. 6: L6–L9
159. Nadeborn W, Andrä P, Osten W (1995) A robust procedure for absolute phase measurement. Opt Lasers Eng 22: 245–260
160. Neal DR, Alford W, Gruetzner JK (1996) Amplitude and phase beam characterization using a two-dimensional wavefront sensor. Proc. SPIE 2870:72–82
161. Neumann DB (1980) Comparative holography. In: Tech digest topical meeting on hologram interferometry and speckle metrology, paper MB2-1. Opt Soc Am, 1764–1766
162. Nilsson B, Carlsson T (1998) Direct three-dimensional shape measurement by digital light-in-flight holography. Appl Opt 37(34):7954–7959
163. Nilsson B, Carlsson T (1999) Digital light-in-flight holography for simultaneous shape and deformation measurement. In: Proc. SPIE 3835: 127–134
164. Nilsson B, Carlsson T (2000) Simultaneous measurement of shape and deformation using digital light-in-flight recording by holography. Opt Eng 39(1):244–253
165. Nimmo Smith W (2008) A submersible three-dimensional particle tracking velocimetry system for flow visualization in the coastal ocean. Limnology & Oceanography: Methods 6:96–104
166. Onural L (2000) Sampling of the diffraction field. Appl Opt 39(32):5929–5935
167. Onural L, Özgen MT (1992) Extraction of three-dimensional object-location information directly from in-line holograms using Wigner analysis. J Opt Soc Am A 9(2):252–260
168. Onural L, Scott PD (1987) Digital decoding of in-line holograms. Opt Eng 26 (11):1124–1132
169. Osten W, Nadeborn W, Andrä P (1996) General hierarchical approach in absolute phase measurement, In: Proc SPIE. vol 2860, 2–13
170. Osten W, Kalms M, Jüptner, Tober G, Bisle W, Scherling D (2000) Shearography system for the testing of large scale aircraft components taking into account noncooperative surfaces, In: Proc SPIE. vol 4101B. 432–8
171. Osten W, Seebacher S, Jüptner W (2001) Application of digital holography for the inspection of microcomponents. Proc SPIE 4400:1–15
172. Osten W, Seebacher S, Baumbach T, Jüptner W (2001) Absolute shape control of microcomponents using digital holography and multiwavelength contouring. Proc SPIE 4275:71–84

173. Osten W, Baumbach T, Seebacher S, Jüptner W (2001) Remote shape control by comparative digital holography. In: Jüptner W, Osten W (eds) Proc 4th International Workshop on Automatic Processing of Fringe Patterns. Akademie, Berlin, pp 373–382

174. Osten W, Baumbach T, Jüptner W (2002) Comparative digital holography. Opt Lett 27 (20):1764–1766

175. Ostrovsky YI, Butosov MM, Ostrovskaja GV (1980) Interferometry by Holography. Springer, New York

176. Owen RB, Zozulya A (2000) In-line digital holographic sensor for monitoring and characterizing marine particulates. Opt Eng 39(8):2187–2197

177. Owen RB, Zozulya A, Benoit MR, Klaus DM (2002) Microgravity materials and life sciences research applications of digital holography. Appl Opt 41(19):3927–3935

178. Pan G, Meng H (2003) Digital holography of particle fields: reconstruction by use of complex amplitudes. App Optics 42:827–833

179. Pedrini G, Tiziani H (2002) Short-coherence digital microscopy by use of a holographic imaging system. Appl Opt 41(22):4489–4496

180. Pedrini G, Zou YL, Tiziani H (1995) Digital double-pulsed holographic interferometry for vibration analysis. J Mod Opt 42(2):367–374

181. Pedrini G, Zou Y, Tiziani H (1997) Simultaneous quantitative evaluation of in-plane and out-of-plane deformations by use of a multidirectional spatial carrier. Appl Opt 36(4):786

182. Pedrini G, Schedin S, Tiziani H (1999) Lensless digital holographic interferometry for the measurement of large objects. Optics Communications 171:29–36

183. Pedrini G, Schedin S, Tiziani H (2000) Spatial filtering in digital holographic microscopy. J Mod Opt 47(8):1447–1454

184. Pedrini G, Titiani HJ, Alexeenko I (2002) Digital-holographic interferometry with an image-intensifier system. Appl Opt 41(4):648

185. Pedrini G, Osten W, Zhang Y (2005) Wave-front reconstruction from a sequence of interferograms recorded at different planes. Opt. Lett. 30:833–835

186. Pech-Pacheco JL, Cristóbal G, Chamorro-Martínez J, Fernández-Valdivia J. (2000) Diatom autofocusing in brightfield microscopy: a comparative study. proceedings of the IEEE international conference on pattern recognition (ICPR'00). 3: 3318–3321

187. Pettersson S-G, Bergstrom H, Abramson N (1989) Light-in-flight recording 6: Experiment with view-time expansion using a skew reference wave. Appl Opt 28:766–770

188. Platt BC, Shack R (2001) History and Principles of Shack-Hartmann Wavefront Sensing. Journal of Refractive Surgery 17:573–577

189. Pomarico J, Schnars U, Hartmann HJ, Jüptner W (1996) Digital recording and numerical reconstruction of holograms: A new method for displaying Light-in-flight. A Opt 34 (35):8095–8099

190. Powell RL, Stetson KA (1965) Interferometric Vibration Analysis by Wavefront reconstructions. J Opt Soc Amer 55:1593–1598

191. Raupach S, Vössing HJ, Curtius J, Borrmann S (2006) Digital crossed-beam holography for in situ imaging of atmospheric ice particles. J. Opt. A: Pure Appl. Opt. 8(9):796

192. Rodenburg JM, Hurst AC, Cullis AG, Dobson BR, Pfeiffer F, Bunk O, David C, Jefimovs K, Johnson I (2007) Hard-X-Ray Lensless Imaging of Extended Objects. Phys. Rev. 98:034801

193. Rolleston R, George N (1986) Image reconstruction from partial Fresnel zone information. Appl. Opt. 25:178–183

194. Santos A, Oritz C, De Soloranzo J, Vaquero J.J., Pena JM,, Malpica N, Del Pozo F et al (1997) Evaluation of autofocus functions in molecular cytogenetic analysis. J Microsc 188 (3): 264–272

195. Schnars U (1994) Direct phase determination in hologram interferometry with use of digitally recorded holograms. J Opt Soc Am A 11(7):2011–2015, reprinted (1997) In: K Hinsch, R Sirohi (eds). SPIE Milestone Series MS 144, pp 661–665

196. Schnars U (1994) Digitale Aufzeichnung and mathematische Rekonstruktion von Hologrammen in der Interferometrie. VDI-Fortschritt-Berichte series 8 no 378 VDI, Düsseldorf
197. Schnars U, Jüptner W (1993) Principles of direct holography for interferometry. In: Jüptner W, Osten W (eds) FRINGE 93 Proc. 2nd International workshop on automatic processing of fringe patterns. Akademie, Berlin, pp 115–120
198. Schnars U, Jüptner W (1994) Direct recording of holograms by a CCD-target and numerical reconstruction. Appl Opt 33(2):179–181
199. Schnars U, Jüptner W (1994) Digital reconstruction of holograms in hologram interferometry and shearography. Appl Optics 33(20):4373-4377, reprinted (1997) In: K Hinsch , R Sirohi (eds). SPIE Milestone Series MS 144, pp 656–660
200. Schnars U, Jüptner W (1995) Digitale Holografie. In: annual conference of the Deutsche Gesellschaft für angewandte Optik. Handout, Binz.
201. Schnars U, Geldmacher J, Hartmann HJ, Jüptner W (1995) Mit digitaler Holografie den Stoßwellen auf der Spur. F&M 103(6):338–341
202. Schnars U, Hartmann HJ, Jüptner W (1995) Digital recording and numerical reconstruction of holograms for nondestructive testing. Proc SPIE 2545:250–253
203. Schnars U, Kreis T, Jüptner W (1996) Digital recording and numerical reconstruction of holograms: Reduction of the spatial frequency spectrum. Opt Eng 35(4):977–982
204. Schreier D (1984) Synthetische Holografie. VCH, Weinheim
205. Schwomma O (1972) austrian patent 298,830
206. Seebacher S (2001) Anwendung der digitalen Holografie bei der 3D-Form- und Verformungsmessung an Komponenten der Mikrosystemtechnik. University Bremen publishing house, Bremen
207. Seebacher S, Osten W, Jüptner W (1998) Measuring shape and deformation of small objects using digital holography. Proc SPIE 3479:104–115
208. Seebacher S, Baumbach T, Osten W, Jüptner W (2000) Combined 3D-shape and deformation analysis of small objects using coherent optical techniques on the basis of digital holography. Proc SPIE 4101B:520–531
209. Seifert L, Tiziani HJ, Osten W (2005) Wavefront reconstruction with the adaptive Shack-Hartmann sensor. Opt. Comm. 245:255–269
210. Sequoia LISSTHOLO. http://www.sequoiasci.com/products/LISSTHOLOspecs.cmsx.
211. Servin M, Malacara D, Marroquin JL (1996) Wave-front recovery from two orthogonal sheared interferograms. Appl. Opt. 35:4343–4348
212. Shack RV, Platt BC (1971) Production and use of a lenticular Hartmann screen (abstract). J. Opt. Soc. Am. 61:656
213. Sheng J, Malkiel E, Katz J (2008) Using digital holographic microscopy for simultaneous measurements of 3D near wall velocity and wall shear stress in a turbulent boundary layer. Exp Fluids 45:1023–1035
214. Sheng J, Malkiel E, Katz J (2006) Digital holographic microscope for measuring three-dimensional particle distributions and motions. App Opt 4:3893–3901
215. Sjoedahl M, Benckert LR (1993) Electronic speckle photography: analysis of an algorithm giving the displacement with subpixel accuracy. Appl Opt 32(13):2278–2284
216. Skarman B, Becker J, Wozniak K (1996) Simultaneous 3D-PIV and temperature measurements using a new CCD-based holographic interferometer. Flow Meas Instrum 7 (1):1–6
217. Snyman J (2005) Practical Mathematical Optimization: An Introduction to Basic Optimization Theory Class New Gradient-Based Algorithms. Springer US, 1 ed.
218. Sollid JE (1969) Holographic interferometry applied to measurements of small static displacements of diffusely reflecting surfaces. Appl Opt 8:1587–1595
219. Steinbichler H (2004) Shearography – NDT. Product information, Steinbichler, Neubeuern
220. Steinchen W, Yang L (2003) Digital Shearography. SPIE press.
221. Stern A, Javadi B (2006) J Opt Soc Amer A: Opt Image Sci 24:163–168

222. Stetson KA, Powell RL (1965) Interferometric hologram evaluation and real-time vibration analysis of diffuse objects. J Opt Soc Amer 55:1694–1695

223. Stetson KA, Brohinsky R (1985) Electrooptic holography and its application to hologram interferometry. Appl Opt 24(21):3631–3637

224. Stetson KA, Brohinsky R (1987) Electrooptic holography system for vibration analysis and nondestructive testing. Opt Eng 26(12):1234–1239

225. Streibl N (1984) Phase imaging by the transport equation of intensity. Opt. Comm. 49:6–10

226. Sun H, Song B, Dong H, Reid B, Player MA, Watson J, Zhao M (2008) Visualization of fast-moving cells in vivo using digital holographic microscopy. Jnl Biomedical Optics 13:014007

227. Sun HY, Hendry DC, Player MA, Watson J (2007) In situ electronic holographic camera for studies of plankton. IEEE J Ocean Eng 32:373–382

228. Sun H, Perkins RG, Watson J, Player MA, Paterson DM (2004) Observations of coastal sediment erosion using in-line holography. J Opt Sci A: Pure and App Optics 6:703–710

229. Sun H, Benzie PW, Burns N, Hendry DC, Player MA, Watson J (2008) Underwater digital holography for studies of marine plankton. Phil Trans Roy Soc 366:1789–1806

230. Synnergren P, Sjödahl M (2000) Mechanical testing using digital speckle photography. Proc SPIE vol 4101B, 520–531

231. Tajahuerce E, Javidi B (2000) Encrypting three-dimensional information with digital holography. Appl Opt 39(35):6595–6601

232. Tajahuerce E, Matoba O, Verral S, Javidi B (2000) Optoelectronic information encryption with phase-shifting interferometry. Appl Opt 39(14):2313–2320

233. Tajahuerce E, Matoba O, Javidi B (2001) Shift-invariant three-dimensional object recognition by means of digital holography. Appl Opt 40(23):3877–3886

234. Takajo H, Takahashi T, Ueda R, Taninaka M (1998) Study on the convergence property of the hybrid input output algorithm used for phase retrieval. J. Opt. Soc. Am. A 15:2849–2861

235. Takaki Y, Ohzu H (1999) Fast numerical reconstruction technique for high-resolution hybrid holographic microscopy. Appl Opt 38(11):2204–2211

236. Takaki Y, Ohzu H (2000) Hybrid holographic microscopy: visualization of three-dimensional object information by use of viewing angles. Appl Opt 39(29):5302–5308

237. Takaki Y, Kawai H, Ohzu H (1999) Hybrid holographic microscopy free of conjugate and zero-order images. Appl Opt 38(23):4990–4996

238. Teague M (1983) Deterministic phase retrieval: a Green's function solution. J. Opt. Soc. Am. A 73:1434–1441

239. Thompson BJ, Ward JH (1966) Particle sizing – the first direct use of holography. Sci Res 1:37–40

240. Thompson BJ (1978) Applications of Holography. Rep. Prog. Phys 41:633–674

241. Trolinger JD (1991) Particle and Flow Field Holography Combustion Measurements. Chigier N (ed). Hemisphere Publishing Corporation, pp 51-89

242. Trolinger J (1975) Particle Field Holography. Optical Engineering 14:383–392

243. Verrier N, Atlan M (2011) Off-axis digital hologram reconstruction: some practical considerations. Appl Opt 50(34):136

244. Vikram CS (1992) Particle field holography, Cambridge

245. Wagner C, Seebacher S, Osten W, Jüptner W (1999) Digital recording and numerical reconstruction of lensless Fourier holograms in optical metrology. Appl Opt 38 (22):4812–4820

246. Wagner C, Osten W, Seebacher S (2000) Direct shape measurement by digital wavefront reconstruction and multiwavelength contouring. Opt Eng 39(1):79–85

247. Waller L, Tian L, Barbastathis G (2010) Transport of Intensity phase-amplitude imaging with higher order intensity derivatives. Opt. Exp. 18:12552–12561

248. Watson J, Alexander S, Craig G, Hendry DC, Hobson PR, Lampitt RS, Marteau JM, Nareid H, Player MA, Saw K, Tipping K (2001) Simultaneous in-line and off-axis subsea holographic recording of plankton and other marine particles. Meas Sci Technol 12:L9–L15

249. Watson J, Burns N (2013) Submersible holography and subsea holocameras, in Subsea Optics and Imaging, J Watson, Z Zielinski (eds), 12: 294–326
250. Winnacker A (1984) Physik von Laser und Maser. BI-Verlag, Mannheim
251. Wozniak K, Skarman B (1994) Digital holography in flow visualization. final report for ESA/ESTEC purchase order 142722, Noordwijk
252. Xiao X, Puri I (2002) Digital recording and numerical reconstruction of holograms: an optical diagnostic for combustion. Appl Opt 41(19):3890–3898
253. Xu L, Peng X, Asundi A, Miao J (2001) Hybrid holographic microscope for interferometric measurement of microstructures. Opt Eng 40(11):2533–2539
254. Yamaguchi I, Saito H (1969) Application of holographic interferometry to the measurement of poisson's ratio. Jap Journal of Appl Phys 8:768–771
255. Yamaguchi I, Zhang T (1997) Phase-shifting digital holography. Optics Letters 22 (16):1268–1270
256. Yamaguchi I, Kato J, Ohta S, Mizuno J (2001) Image formation in phase-shifting digital holography and applications to microscopy. Appl Opt 40(34):6177–6186
257. Yamaguchi I, Inomoto O, Kato J (2001) Surface shape measurement by phase shifting digital holography. In: Jüptner W, Osten W (eds) Proc 4th International Workshop on Automatic Processing of Fringe Patterns. Akademie, Berlin, pp 365–372
258. Yamaguchi I, Matsumura T, Kato J (2002) Phase-shifting color digital holography. Opt Lett 27(13):1108–1110
259. Yang S, Xie X, Thuo Y, Jia C (1999) Reconstruction of near-field in-line holograms. Optics Communications 159:29–31
260. Yang G, Dong B, Gu B, Zhuang J, Ersoy OK (1994) Gerchberg-Saxton and Yang-Gu algorithms for phase retrieval in a nonunitary transform system: a comparison. Appl. Opt. 33:209–218
261. Yaroslavskii LP, Merzlyakov NS (1980) Methods of digital holography. Consultants Bureau, New York
262. Youla DC, Webb H (1982) Image restoration by the method of convex projections: Part 1 – Theory. IEEE Trans. Med. Imaging MI-1: 81–94.
263. Yu L, Cai L (2001) Iterative algorithm with a constraint condition for numerical reconstruction of a three-dimensional object from its hologram. J Opt Soc Am A 18 (5):1033–1045
264. Zhang T, Yamaguchi I (1998) Three-dimensional microscopy with phase-shifting digital holography. Optics Letters 23(15):1221–1223
265. Zhang T, Yamaguchi I (1998) 3D microscopy with phase-shifting digital holography. Proc SPIE 3479:152–159
266. Agour M, Huke P, v. Kopylow C, Falldorf C (2010) Shape measurement by means of phase retrieval using a spatial light modulator. In: AIP Conf. Proc. 1236:265–270
267. Agour M (2012) Determination of the complex amplitude of monochromatic light from a set of intensity observations. Strahltechnik 47, BIAS Verlag Bremen
268. Lord Rayleigh, Wood R W (1898) Phase reversal zone plates and diffraction telescope. Phil. Mag. Series 5, vol. 45: 511. Reprinted in Lord Rayleigh Scientific Papers. pp. 74–79 (1887–1892)
269. Sutkowski M, Kujawinska M (2000) Application of liquid crystal (LC) devices for optoelectronic reconstruction of digitally stored holograms. Optics and Lasers in Engineering 33(3):191–201

Index

Symbol
4f-setup, 147

A
Aberrations, 104, 196
Aircraft industry, 70
Amplitude, 6, 21, 22
Amplitude transmission, 22
Amplitude transmittance, 22
Angular frequency, 6
Angular spectrum method, 49
Aperture, 48, 185
Aperture size, 137, 138
Ar-Ion laser, 121
Astigmatism, 116
Autocorrelation function, 15, 123
Autofocusing, 118

C
Charged coupled device (CCD), 39, 58
Circle of confusion, 138
CMOS, 39, 58, 99
Coherence, 2, 10, 66
Coherence distance, 14, 15
Coherence length, 12, 66, 121, 123
Coma, 116
Comparative digital holography, 4, 131, 132
Comparative interferometry, 132
Complex amplitude, 8, 17, 21, 41, 57
Complex degree of coherence, 15
Computational shear interferometry, 142, 177, 180
Computational wave field sensing, 141
Computational wave front sensing, 2
Computer generated holography, 2
Conjugate image, 98
Conjugate object wave, 23

Conjugate reference, 41
Constructive interference, 9
Contouring, 30
Contrast, 19, 63, 66
Convolution approach, 49, 51
Convolution theorem, 50
Correction factor, 42, 51
CoSI, 177
Cross correlation, 15
Cross correlation function, 193

D
DC term, 54, 65
Decorrelation, 132
Demodulation, 37, 38
Depth of focus, 1, 21, 102, 160
Depth-of-field, 96, 102
Destructive interference, 9
Deterministic methods, 156
Diffraction, 5, 15
Diffraction efficiency, 131
Diffuser, 94
Digital Fourier holography, 52
Digital hologram, 46, 132
Digital holographic interferometry, 41, 69, 194
Digital holographic microscopy, 4, 106, 107
Digital holography, 2, 69, 124, 127
Digital mirror device, 130
Digital speckle pattern interferometry, 185
Digital speckle photography, 193
Diode laser, 12
Displacement vector, 28, 29, 70, 188
Displacement vector field, 74
Double-exposure holography, 26
Dye laser, 121
Dynamic evaluation, 84
Dynamic range, 63

© Springer-Verlag Berlin Heidelberg 2015
U. Schnars et al., *Digital Holography and Wavefront Sensing*,
DOI 10.1007/978-3-662-44693-5

E

Eddy current, 82
Electrical field, 6
Electromagnetic wave, 5
Electronic speckle pattern interferometry, 3,
    185, 194
Electro-optic holography, 189
Encrypting of information, 4
Endoscopic digital holography, 127
Error reduction method, 150

F

Fizeau interferometer, 176
Flaws, 83
Focal distance, 41
Fourier hologram, 2
Fourier transform, 50, 53
Fourier holography, 00000
Fourier transform method, 36
Fourier transformation, 44
Fractional Fourier transform, 147
Frame grabber, 127
Frame-transfer architecture, 61
Fraunhofer far-field distance, 98
Fraunhofer model, 96
Frequency, 6
Fresnel approximation, 42
Fresnel hologram, 2, 97
Fresnel lens, 1
Fresnel transform, 49, 70
Fresnel transformation, 43
Fresnel-Kirchhoff integral, 16, 17, 39, 49, 57
Fringe, 9
Full-frame architecture, 61

G

Gerchberg-Saxton approach, 144
Gerchberg-Saxton scheme, 146
Glass fibres, 127
Gradient search methods, 154, 170
Grating, 126

H

Helmholtz-equation, 142, 179
Heterodyne, 36
Hierarchical phase unwrapping, 90
Hilbert space, 145
Holocameras, 96, 115
Hologram, 21, 39
Holographic interferogram, 27, 69, 130, 187
Holographic interferometry, 2, 25, 35, 194

Holographic microscopy, 95
Holographic particle image velocimetry, 99
Holography, 1, 20, 46
Holovideos, 105, 117
Huygens' principle, 16, 17
Hybrid input output method, 150, 152

I

Illumination direction, 74
Image plane, 113
Image plane holograms, 196
Imaging equations, 23
Impact loading, 70
Impulse response function, 50
Inclination factor, 17
Incoherent light, 13
Information encryption, 135
In-line, 96, 97
In-line holography, 2
In-plane, 187, 193
Intensity, 7, 19, 22, 43
Interference, 1, 8
Interference pattern, 1
Interference phase, 27, 30, 37, 69, 70, 83, 188,
    196, 197
Interferogram, 69
Interferometer, 122
Interline-transfer architecture, 60
Inverse problem, 141
Inversion, 58

L

Laplace operator, 5
Laser, 12, 20, 121
Laser diode, 128
Lateral magnification, 25
LCD, 173, 175- 177
LED, 114, 149, 160, 182
Lens array, 183
Lens transmission factor, 51
Light emitting diode, 68
Light source, 66
Light-in-flight holography, 4, 122
Linear polarized light, 6
Liquid crystal spatial light modulator, 147, 175
Longitudinal magnification, 25

M

Mach-Zehnder interferometer, 92, 113
Magnetic field, 5
Magnification, 25, 51, 108

Maxwell equations, 5, 7
Michelson interferometer, 11, 126, 191
Monochromatic, 8, 12
Multiwavelength contouring, 87

N

Natural extension, 167
Non-destructive testing, 81
Numerical focussing, 196
Numerical hologram reconstruction, 2
Numerical reconstruction, 42, 124
Numerical refocusing, 114

O

Object wave, 1, 23
Objective function, 143, 173, 179
Objective speckle pattern, 19
Observation direction, 74, 86
Off-axis, 46
Optical fibres, 85
Optical path difference, 66, 121
Optical Reconstruction, 129
Orthoscopic image, 25
Out-of plane, 187
Out-of-plane deformation, 30

P

Partially coherent light, 13
Particle image velocimetry, 106, 114
Particle measurement, 2
Particle sizing, 96
Particle tracking, 4
Penetrant testing, 82
Perspective, 1, 21
Phase, 7, 21, 22
Phase aberrations, 42
Phase object, 34
Phase retrieval, 142, 148
Phase shift, 113, 128
Phase shift angle, 118
Phase shifting, 188, 196
Phase shifting digital holography, 4, 56
Phase shifting holographic interferometry, 35
Phase shifting interferometry, 180- 182
Phase singularities, 164
Phase unwrapping, 37, 70
Photo effect, 59
Photographic emulsions, 62, 63
Photographic plates, 20, 196
Photons, 5
Photorefractive crystals, 195

Phytoplankton, 116
Piezoelectric transducer, 56, 113
Piezoelectric translator, 36
Pinhole camera effect, 138
Pixel distance, 47
Plane reference wave, 41
Plane wave, 6
Plane wave decomposition, 149
Plankton, 118
Poisson noise, 169
Poisson ratio, 74
Printer, 131
Propagation operator, 143
Pseudoscopic, 99
Pseudoscopic image, 25
Ptychography, 162

R

Real image, 23, 39, 130
Real time technique, 27
Reconstruction, 39
Reference wave, 1, 21, 39, 124
Refractive index, 34, 92
Resolution, 49, 51, 62, 196
Rigid body motions, 81, 191
Ruby laser, 73

S

Sensitivity, 63
Sensitivity vector, 30, 74
Shack-Hartmann, 4, 183, 184
Shape Measurement, 85
Shear interferometry, 4, 162
Shear-interferometer, 175
Shearogram, 190
Shearography, 3, 82, 189
Shutter, 60
Skeletonizing, 36
Spatial coherence, 13, 114
Spatial frequency, 10, 20, 62, 65, 72
Spatial light modulator, 129
Speckle, 18, 185
Speckle decorrelation, 88
Speckle interferogram, 186, 190
Speckle interferometry, 177
Speckle pattern, 193
Speckle photography, 3
Speckle size, 19, 47, 185, 187
Spectral width, 12
Speed of light, 5, 125
Spherical aberration, 116
Spherical reference wave, 53

Stability, 66
Steepest descent gradient method, 154, 171
Strains, 193
Subjective speckle pattern, 20
Submersible, 115
Subpixel evaluation, 194
Summed distance error, 146
Superluminescent diode, 67
Superposition, 8
Suppression, 53
Synthetic apertures, 137
Synthetic wavelength, 32, 87, 90

T

Telecentric imaging system, 34
Temporal coherence, 11
Temporal phase unwrapping, 128
Thermal expansion coefficient, 74, 78
Thermoplastic films, 195
Tilted reference wave, 56
Tomography, 110, 126
Torsions, 75
Transient deformations, 72
Transparent media, 92
Transport-of-intensity equation, 156
TV-holography, 185
Twin image, 2, 93
Two-illumination-point method, 30, 85
Two-wavelength contouring, 134
Two-wavelength method, 30, 86

U

Ultrasonic testing, 82
Underwater, 115
Unwrapped phase, 73

V

Vacuum chamber, 78
Vacuum permittivity, 7
Van Cittert-Zernike theorem, 173
Vibration isolation, 189, 193
Vibrations, 66, 128
Virtual image, 21, 23, 39, 55, 130
Virtual lens, 42
Visibility, 12

W

Wave equation, 5
Wave field reconstruction, 178
Wave front reconstruction, 164, 178
Wave number, 6
Wave vector, 6
Wavefield sensing, 4
Wavefront, 9, 16, 122, 125
Wavelength, 6

X

X-ray, 82

Y

Young interferometer, 13, 15
Young's interferometer, 14
Young's modulus, 74, 76

Z

Zero order, 46, 54, 128
Zero padding, 63